Microbial Systematics

Taxonomy, Microbial Ecology, Diversity

T0186479

Editor

Bhagwan Rekadwad, Ph.D.

National Centre for Microbial Resource
NATIONAL CENTRE FOR CELL SCIENCE
NCCS Complex, Ganeshkhind Road, Pune 411007
India

Institute of Bioinformatics and Biotechnology
SAVITRIBAI PHULE PUNE UNIVERSITY
Ganeshkhind, Pune 411007
Maharashtra, India

CRC Press
Taylor & Francis Group
Boca Raton   London   New York

CRC Press is an imprint of the
Taylor & Francis Group, an **informa** business

A SCIENCE PUBLISHERS BOOK

Cover credit: Cover illustrations reproduced by kind courtesy of the editor, Dr. Bhagwan Rekadwad.

CRC Press
Taylor & Francis Group
6000 Broken Sound Parkway NW, Suite 300
Boca Raton, FL 33487-2742

First issued in paperback 2022

© 2021 by Taylor & Francis Group, LLC
CRC Press is an imprint of Taylor & Francis Group, an Informa business

No claim to original U.S. Government works

Version Date: 20200609

ISBN-13: 978-0-367-64649-3 (pbk)
ISBN-13: 978-0-367-14801-0 (hbk)

DOI: 10.1201/9780429053535

Library of Congress Cataloging-in-Publication Data

Names: Rekadwad, Bhagwan, 1983- editor.
Title: Microbial systematics : taxonomy, microbial ecology, diversity /
   editor, Bhagwan Rekadwad, National Centre for Microbial Resource,
   National Centre for Cell Science, Pune, Maharashtra, India.
Description: Boca Raton, FL : CRC Press, [2020] | Includes bibliographical
   references and index.
Identifiers: LCCN 2020019401 | ISBN 9780367148010 (hardcover)
Subjects: LCSH: Microbiology. | Microbiology--Classification. | Microbial
   ecology. | Bacterial diversity.
Classification: LCC QR41.2 .M463 2020 | DDC 579--dc23
LC record available at https://lccn.loc.gov/2020019401

Preface

Researchers studying "Microbial Systematics" are few as compared to the other specializations in Natural Sciences. I have long felt the need for an updated text covering the basic and applied approaches for microbial classification, identification, taxonomy, diversity, ecology, applications. Science is progressing at a fast pace and new technologies are rapidly emerging. The use of basic methodologies and new-generation sequencing techniques for evolutionary analysis of microorganisms can be exploited consistently to generate vast amount of information. Parallels to generation, and use of new technologies, microorganisms are used for various applications. There is an urgent need to have a volume, which includes earlier research. This book provides a paradigm shift for the systematics community. It has also uncovered other facets of microbial systematics including microbiome and their societal applications. Moreover, systematic studies on microorganisms have been exploited for their re-classification and applications based on the identification of genes, ability to produce macromolecules, and their uses in research and industries. It inspires scientists in the field of microbial ecology to explore microorganisms, soil microbiota, human and animal microbiome, patterns of survival and secretions of biomolecules, which is a key for today's researchers and future generation of scientists. This book came into existence with the enormous support from authors and experts. I am most grateful to the author(s) of the book chapters who have generously provided their research literature and shared views to shape this book. I am thankful to Dr. Juan M. Gonzalez, Microbial Diversity and Microbiology of Extreme Environments Research Group, Agencia Estatal Consejo Superior de Investigaciones Científicas, IRNAS-CSIC, Sevilla, Spain for explaining his views about Microbial Systematics in real time, Prof. Wen-Jun Li, an internationally renowned scientist in microbial systematics and microbial ecology, State Key Laboratory of Biocontrol and Guangdong Provincial Key Laboratory of Plant Resources, School of Life Sciences, Sun Yat-Sen University, Guangzhou, PR China for his contribution in the form of book chapter, I am indebted to Dr. Yogesh S. Shouche, Scientist "G" and Principal Investigator, National Centre for Microbial Resource, National Centre for Cell Science, Pune, India for critical evaluation of research and fair judgments time-to-time, Dr. Vipin Chandra Kalia, Chief Scientist, CSIR-Institute of Genomics and Integrative Biology, Delhi, Scientist at Konkuk University, Seoul, Republic of Korea for his moral support, I am grateful to Professor William B. Whitman, University of Georgia, USA and Professor Brian P. Hedlund, University of Nevada, Las Vegas, USA for their encouragement and motivation during series of BISMiS meetings, and words are insufficient to express my sincere gratitude to my mentor Dr. Kamlesh Jangid, Scientist, National Centre for Microbial Resource (NCMR), National Centre for Cell Science, Pune, India, President-elect at BISMiS—Bergey's International Society for Microbial Systematics, Associate Chief Editor at Soil Biology & Biochemistry, Elsevier for an opportunity to initiate research for his guidance in Microbial Ecology, Taxonomy and Diversity and Microbiome.

April 2020   **Bhagwan Rekadwad**

Contents

Chapter 1

Culture-Dependent and -Independent Strategies in Bacterial Diversity Appraisal

Meora Rajeev, T. J. Sushmitha and
*Shunmugiah Karutha Pandian**

Introduction

The biosphere is dominated by microorganisms and expected to contain $4\text{–}6 \times 10^{30}$ prokaryotic cells (Whitman et al. 1998, Sogin et al. 2006). Microbes are highly diverse and account for 60% of Earth's total biomass. Their abundance correspond to more than three orders of magnitude than cumulative of all plants and animals (Singh et al. 2009). Though invisible to unaided eye, they are ubiquitous and conspicuous in natural environments as they are key players in various activities such as maintenance of carbon dynamics, soil structure formation, decomposition of xenobiotic and organic matters, biogeochemical cycling, food web, recycling of nutrients, and maintenance of ecosystem. They are promising candidates and rich sources for many drugs, bioactive compounds and enzymes.

Three fundamental questions that arise modern ecology are: (i) Who are they? (ii) What do they do? (iii) How their activities influence the ecosystem functions? Microbial ecologists intends to answer these fundamental questions and deals with the study of microbes, their interactions with each other and with their surrounding environment.

Microbial diversity is a general term used to include number of different species (richness) and their relative abundance (evenness) in a particular habitat. Analysis of bacterial community composition and understanding their spatio-temporal patterns has become a longstanding ecological concern. At times, when the importance of bacteria was seizing the limelight, many techniques were developed to study and

Department of Biotechnology, Alagappa University, Science Campus, Karaikudi, 630003, Tamil Nadu, India.
* Corresponding author: sk_pandian@rediffmail.com

identify the diverse nature of bacteria. Microbiologists believed that every single bacterial community has a unique role in maintaining the balance of ecosystem and they form the foundation of the Earth's biosphere.

Indisputably, the discovery of microorganisms by Antonie van Leeuwenhoek and his observation on *Vorticella* beating its cilia developed various approaches to study animalcules (microorganisms, as named by Leeuwenhoek). Robert Koch, a successor in this field, introduced a new method to produce pure culture (homogenous population of microbial cells derived from a single species), postulated them and hired the use of solid media with coagulated egg albumen, starch paste and nutrient agar to cultivate microorganisms. In the year 1882, the age of culturing bacteria on agar was established with Fannie Hesse, who suggested Koch to replace gelatin with agar. However, there has been discrepancy between the culturable bacteria on agar plate and total cell count, an enigma often termed as "great plate count anomaly" (Fig. 1), which was the central turn in microbiology. Although bacterial systematics was widened with biochemical characterization (proposed in Bergey's Manual) such as fatty acid methyl esters (FAME) analysis, enzyme production tests, carbon source utilization profile, community level physiological profile and many specified analysis for particular group of bacteria, even then it was not possible to analyze whole bacterial community. Subsequently in the early 20th century, bacterial systematics was widely propagated and many researchers provided new insights into bacterial cellular structure, growth conditions and biochemical pathways through technical advancements. These eminent microbiologists established a strong foundation in understanding the importance of bacteria in ecosystem functioning and paved the way to study their physiology, genetics and functions.

All organisms present in an ecosystem depend on microbial activity. On the other hand, the abundant and diverse microorganisms are both determinants and

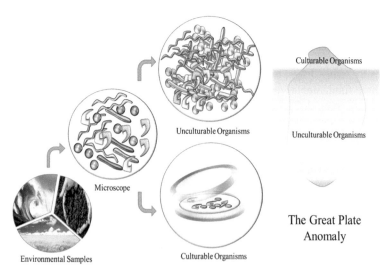

Fig. 1. Great Plate Count Anomaly: Staley and Konopka (1985) observed a large disparity between the viable plate count and direct microscopic count of bacterial community present in samples of various habitats. Direct microscopic counting was several order of magnitude higher than the plating count.

indicators of healthy ecosystem. Before discussing how variation in structure of bacterial community influences the ecosystem functions, there is a demand for reliable and accurate methods for studying environmental microbes to determine their phylogenetic position and relative proportion.

Progress in bacterial identification has been accelerating since decades and has resulted in molecular insights. Bacterial identification and diversity appraisal with DNA and RNA has enlarged its visualization, through which many technological breakthroughs and revolutions have occurred. Upraise in employing of 16S rRNA gene as a barcode for bacterial identification in several techniques from culture-dependent identification through Sanger sequencing to culture-independent methods such as DGGE, ARDRA, T-RFLP and next-generation sequencing (NGS) is proof for technology development and the knowledge we have gained from the last two to three decades. The thirst for knowledge on bacterial diversity and community structure in various habitats led to the launch of various megaprojects such as Human Microbiome Project (HMP) and Earth Microbiome Project (EMP) to improve our understanding of microbial flora involved in human health as well as to characterize microbial pattern on Earth (Gilbert et al. 2014). The progress made during the past few years directly reflects the importance of bacterial diversity on the Earth planet. With this background, this chapter deals with the major techniques used for assessment of bacterial diversity in various environments, its principles and comparison.

Ecological perception of bacterial diversity

Understanding the bacterial diversity in various environments is an important aspect as bacteria occupies the majority of the Earth's species diversity and serve as engine in carrying out several ecologically important processes such as recycling of nutrients, remineralisation, degradation of pollutants and decomposition of organic matters. In all ecosystems (particularly marine environment), heterotrophic bacteria together with other microorganisms such as fungi, viruses and microalgae produce enzymes that degrade complex polymers into low-molecular weight substances and facilitate the turnover of various organic compounds. Indeed, these ecological processes eventually sustain lives on Earth.

In addition, analysis of bacterial diversity is also important from the prospect of ecosystem functioning as it determines the ecosystem stability including resistance against anthropogenic activities and resilience. Therefore, establishment of a correlation between biodiversity and ecosystem functioning is a major scientific quest in environmental studies (Loreau et al. 2001).

Ribosomal DNA based bacterial identification

Gene coding for 16S ribosomal RNA 16S rRNA gene or 16S rDNA remains the versatile and standard molecular marker for culture-dependent and -independent profiling of microbial diversity. In bacterial and archaeal cells, 16S rDNA is part of small subunit ribosomal RNA (SSU rRNA) and has proven to be an invaluable tool for phylogenetic studies as this gene is universally present across the members of bacterial domain, is highly conserved and has low evolutionary rate (Nguyen et al. 2016). Though it is highly conserved, the hypervariable regions with noticeable

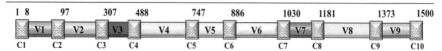

Fig. 2. Schematic depiction of hypervariable regions of bacterial 16S rRNA gene displaying conserved regions (C1–C10) and hypervariable regions (V1–V9). The number above the gene denotes start nucleotide (*E. coli* position) of each hypervariable region (Reproduced from Yang et al. 2016. Sensitivity and correlation of hypervariable regions in 16S rRNA genes in phylogenetic analysis. BMC Bioinformatics 17(1): 135).

sequence diversity present in the rDNA can be used to distinguish bacterial species (Fig. 2). These nine hypervariable regions (V1–V9) are flanked by conserved regions (C1–C10) which can serve as template for primer binding in PCR amplification (Ward et al. 1990). Identification of bacteria includes the amplification of 16S rRNA gene using extracted genomic DNA as template followed by sequencing. The obtained sequence is compared with known sequences in public databases to identify its close relative.

In culture-dependent methods, Sanger sequencing allow high quality sequencing of about 1,000 nucleotides and revolve around nearly full-length sequencing of 16S rDNA. However, various methods such as DGGE, TGGE and high-throughput next-generation sequencing technologies are optimized for short read length (350–500 bp in length) and thus utilize either a single or a combination of two or three of nine hypervariable regions (Li et al. 2009). All frequently used universal primers that are designed to target various hypervariable regions of 16S rRNA gene are conceived to demonstrate different coverage for different bacterial taxa. Both computational and empirical research have revealed that selection of primer pairs encompassing different hypervariable regions could have a significant impact on determination of bacterial diversity and therefore are a critical determinant of diversity assessment.

Cai et al. (2013) conducted a comprehensive study to decipher the effect of choice of primer sets on obtained results of bacterial community structure. This study compared amplification products of primers targeting V_1–V_2, V_3–V_4, V_5–V_6, V_7–V_9, V_1–V_4, V_3–V_6 and V_5–V_9 hyper-variable regions of 16S rRNA gene and suggested that V_3–V_4 regions performed well with little bias and high coverage.

Apart from this, several other problems such as inhibition of PCR amplification by co-extracted contaminants, formation of artefactual PCR products, differential amplification and 16S ribosomal RNA gene heterogeneity would unescapably lead to erroneous reflection of microbial diversity, if PCR is employed to amplify 16S rDNA from environmental communities.

Culture-dependent methods in microbial ecology

Though culture-dependent techniques cover only 1% of total bacterial community, they represent the actual live and viable cells of the environmental samples. Culturing of bacteria is indispensable to increase our understanding towards behavior of an organism. Culture-independent techniques cannot substitute culture-dependent techniques as physical, chemical and functional characteristics of a bacterium largely

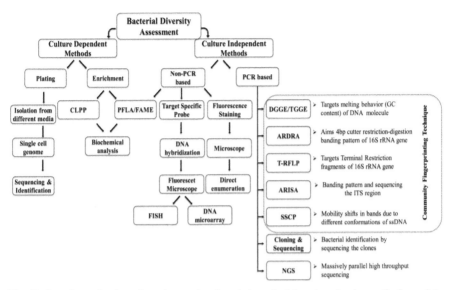

Fig. 3. Overview of culture-dependent and culture-independent based molecular methods used to characterize bacterial diversity in the environment.

rely on well-annotated genome obtained from its cultivated state. Culture-dependent bacterial diversity analysis mainly includes plate count techniques, fatty acid methyl ester analysis and community level physiological profiling through enrichment methods (Fig. 3).

Viable plate count and direct microscopic count

Traditional findings on bacterial structure and its function relied on culturing them on agar plates and in liquid medium. Various culturing methods were designed to maximize the yield of bacteria and to detect diverse number of species. Plating targets a large group of viable, fast growing heterotrophic bacterial cells from which pure culture of individual bacterial species can be studied.

The protocol in general follows making a suspension of known concentration of sample (e.g., soil, water, tissue) in a known volume of solvent and diluting them to a known degree to reach countable number of bacterial colonies. Bacterial counts in diluted samples are estimated by spread plating a known volume (generally 100–200 μl) on different growth media and incubating the petri plates for overnight to days. Diversity of bacteria can be directly visualized in a cultured petri plate where the colony characteristics such as shape, pigment production, colony form (elevated, filamentous, circular), colony surface and opacity (shiny, smooth, wrinkled) and sometimes even the smell of particular colony can give the basic information about diversified nature of the bacterial community. In this method, bacterial count (colony forming unit ml^{-1}) is described by the ratio of number of colonies to volume added (in ml) during plating multiplied to the dilution factor.

Microscopic and staining techniques provide 10–1,000 times higher diversity than the plating methods as these methods involve visualization of samples direct from the environment and do not require any growth condition. Advanced microscopic technique such as fluorescence microscopy utilizes different dyes that particularly targets proteins, lipids and nucleic acids including fluorescein isothiocyanate (FITC), ethidium bromide (EtBr), and acridine orange (AO) and few more to differentiate live and dead cells. However, covering all the millions of diverse bacteria under single lens is impracticable to determine the actual bacterial diversity in an environment.

Phospholipid fatty acid (PLFA) analysis

This technique relies on the signature functional group of fatty acid present in bacteria that facilitates bacterial adaptation in an environment. Proportion of phospholipid fatty acids (PLFAs) differs among bacterial taxonomic groups and therefore used as a marker for bacterial diversity and community interpretation. This technique practices fatty acid methyl esters (FAMEs) as an indicative and taxonomic discriminator for bacterial identification. Phospholipids fatty acids are key components of microbial cell membrane as they form an intact structure in live cell with fatty acid as major component. The chain lengths and composition of these signature PLFAs can determine the type of bacterial community present in the given habitat.

The standard protocol involves four key steps: (i) Extraction of fatty acids from the source samples with the aid of solvent(s) (ii) fractionation of fatty acids by solid phase extraction using column to separate phospholipids from other lipids (iii) saponification followed by methylation of phosphor fatty acids to produce FAMEs (iv) quantitative analysis of FAMEs by employing gas chromatography mass spectrometry (GC-MS). PLFA analysis has been extensively employed in analysis of soil bacterial community and their variations as compared to other habitats. Researchers have assessed the bacterial community changes with agricultural practices (Zelles et al. 1992, 1995, Reichardt et al. 1997), change in season and soil quality (Bossio et al. 1998), rhizosphere community (Garland 1996) and soil exposed to heavy metal pollution (Frostegård et al. 1993).

Regardless of the practicality of this method, it has major limitations (Haack et al. 1995) such as indicative signature fatty acid or phospholipid biochemistry of all bacterial phyla is not available and in certain cases, the signature fatty acid does not necessarily define the particular bacterial community. PLFA cannot categorize the organism to species level. The presence of dominant fatty acid group in distinct bacterial communities misinterprets the bacterial community composition of an environment. Importantly, changes in fatty acid composition occur with respect to surrounding environmental factors and nutrient availability (Šajbidor 1997) that gives a false community estimation. Finally, the difficulty in accessing database and literature of fatty acids of all bacterial community marks a major drawback.

Community-level physiological profiles (CLPP)

Otherwise known as BIOLOG® approach, this aids in measuring the microbial functional diversity by utilizing different forms of sole carbon source. BIOLOG® is a microtiter plate designed for community-level physiological profiles that classifies bacteria based on their capability to oxidize 95 different carbon sources (Table 1)

Table 1. Carbon sources in BIOLOG® microtiter plates.

Carbohydrates	Esters	Carboxylic acids	Amino acids
N-Acetyl-D-galactosamine	Mono-methylsuccinate	Acetic acid	D-Alanine
N-Acetyl-D-glucosamine	Methylpyruvate	cis-Aconitic acid	L-Alanine
Adonitol		Citric acid	L-Alanyl-glycine
L-Arabinose	**Polymers**	Formic acid	L-Asparagine
D-Arabitol	Glycogen	D-Galactonic acid lactone	L-Aspartic acid
Cellobiose	α-Cyclodextrin	D-Galacturonic acid	L-Glutamic acid
i-Erythritol	Dextrin	D-Gluconic acid	Glycyl-L-aspartic acid
D-Fructose	Tween 80	D-Glucosaminic acid	Glycyl-L-glutamic acid
L-Fucose	Tween 40	D-Glucoronic acid	L-Histidine
D-Galactose		α-Hydroxybutyric acid	Hydroxy-L-proline
Gentiobiose	**Brominated chemicals**	β-Hydroxybutyric acid	L-Leucine
α-D-Glucose	Bromosuccinic acid	γ-Hydroxybutyric acid	L-Ornithine
m-Inositol		Itaconic acid	L-Phenylalanine
α-Lactose	**Amides**	α-Ketobutyric acid	L-Proline
Lactulose	Succinamic acid	α-ketoglutaric acid	L-Pyroglutamic acid
Maltose	Glucuronamide	α-Ketovaleric acid	D-Serine
D-Mannitol	Alaninamide	D,L-Lactic acid	L-Serine
D-Mannose		Malonic acid	L-Threonine
D-Melibiose	**Phosphorylated chemicals**	Propionic acid	D,L-Carnitine
β-Methylglucoside	D,L-α-Glycerol phosphate	Quinic acid	γ-Aminobutyric acid
Psicose	Glucose-1-phosphate	D-Saccharic acid	
D-Raffinose	Glucose-6-phosphate	Sebacic acid	**Amines**
L-Rhamnose		Succinic acid	Phenylethylamine
D-Sorbitol	**Aromatic chemicals**	ρ-Hydroxyphenylacetic acid	2-Aminoethanol
Sucrose	Inosine		Putrescine
D-Trehalose	Urocanic acid	**Alcohols**	
Turanose	Thymidine	2,3-Butanediol	
Xylitol	Uridine	Glycerol	

(Adapted from Garland and Mills'. 1991. Classification and characterization of heterotrophic microbial communities on the basis of patterns of community-level sole-carbon-source utilization. Applied and Environmental Microbiology 57(8): 2351–2359).

causing reduction in tetrazolium violet dye. The redox reaction instigated due to bacterial respiration activity reasons a change in color of the medium, which is measured spectrophotometrically. Garland and Mills (1991) initially developed this technique by employing commercially available 96 microtiter plates and differentiated bacteria through sole source carbon utilization (SSCU) pattern. Separate plates are available for Gram-positive and Gram-negative bacteria.

The protocol briefly involves the enrichment of bacteria in a growth medium, followed by addition of equal number of cells in each well. The well containing tetrazolium violet dye generates color as bacteria start utilizing the available carbon source resulting in a metabolic fingerprint (Bochner 1989). In this protocol, utilization of each carbon source denotes a specific bacterial community that need to be compared with the database of BIOLOG®.

CLPP by BIOLOG® emerged as a promising tool in bacterial function diversity assessment that showed high speed, simplified with efficient and standardized protocol. As compared to other techniques, this technique drastically reduced the time needed to assess bacterial diversity of an environment. However, several limitations of this technique such as preference of fast growing bacteria, exact incubation time, equal density of microbial cells/equal inoculum size and its data analysis have also been reported (Lladó and Baldrian 2017).

CLPP has been successfully employed in analyzing bacterial diversity of contaminated soils (Derry et al. 1998), plant rhizospheres (Ellis et al. 1995, Garland 1996), artic soils and soil environment (Derry et al. 1999, Lladó and Baldrian 2017), monitoring of compost maturity (Mondini and Insam 2003) and freshwater community (Christian and Lind 2006).

Culture-based methods in microbial ecology: limitations

As culture-dependent assessment favors only fraction of microbial community, these methods failed to provide information on whole bacterial community present in an environmental sample. The "great plate count anomaly" by Staley and Konopka (1985) clearly proved that the diversity of bacteria that is observed under microscope and grown in petri dishes shows vast disparity. The major problem with culture-dependent method arises from the fact that artificial medium in any laboratory supports the growth of only a fraction of total organisms. The reason for divided growth may be surmised by the fact that: (i) laboratory conditions failing to support ecological niches that are encountered in complex environment (ii) as replication rate varies for bacterial species, the culturable fraction deforms during growth because fast growing species outcompete the slow growing ones (iii) growth requirements such as nutrients and their concentrations, optimum pH, osmotic conditions and temperature of all microbial species are unknown and unpredictable (iv) interdependency of bacteria on their host or each other, viz., co-culturing and endosymbiotic relationship cannot be provided in the laboratory conditions.

The diversity of bacteria with its unique and expanded biosynthetic pathway leading to differentiated growth conditions and parameters restricted researchers to identify whole bacterial community present in a particular environment. With increase in knowledge and understanding of bacterial ecology, various methodological improvements were adapted. These developments include the simulation of environmental parameters at laboratory conditions to increase the chance of cultivation. However, at the end, only 1% of the total microbial community is cultivable in any environment.

Culture-independent/molecular methods for microbial diversity analyses

Culture-dependent epoch gives a huge collection of bacteria and provides information on genetics, physiology and evolution of grown bacteria. However, huge disparity between cultured and actual diversity has increased the importance of culture-independent methods. In addition to this, culturing takes several days for analysis. The utilization of advanced techniques and molecular approaches (Fig. 3) publicized that the bacterial diversity is far superior than predicted until now. The diversity through culture-independent analysis involves PCR based methods (DGGE, T-RFLP, ARISA, ARDRA and SSCP) and non-PCR based methods (Fluorescence *In Situ* Hybridization (FISH) and DNA microarray). In general, these techniques include isolation of metagenome (genome of total microbiota) directly from the environmental source, amplification of molecular marker and finally the sequencing of nucleic acid that determines the bacterial diversity. The principal target for assessing the bacterial diversity has been 16S rRNA gene. Other functional genes such as one coding RNA polymerase B (*rpo B*), methane mono-oxygenase (*mmo A*), nitrogenase (*nif H*), nitrite reductase (*nir S/nir K*) and ammonium mono-oxygenase (*amo A*) are effectively employed as markers to delineate bacterial diversity (Nocker et al. 2007).

PCR based techniques

The advent of polymerase chain reaction (PCR) by Mullis et al. (1986) led to the rapid and efficient improvement in molecular studies. Recombinant DNA technology and genetic engineering has PCR and cloning as their core processes. PCR targeting molecular markers such as prokaryotic 16S rDNA, eukaryotic 18S rDNA and internal transcribed spacer (ITS) region have been extensively studied for identification and prediction of phylogenetic relationship and therefore to explore the yet uncultured bacterial community. The major culture-independent techniques that rely on PCR are clone library, DGGE, T-RFLP, ARDRA, ARISA and SSCP.

Clone library

The most suitable method to analyze bacterial diversity is to amplify the 16S rRNA gene from an environmental sample and then to determine the sequence of individual gene (DeSantis et al. 2007). Obtained nucleotide sequences are compared with known sequences in public database and allocated to most relative taxonomic group.

First step in this method involves amplification of nearly full-length 16S rRNA gene from metagenomic DNA of an environmental sample. Amplified and purified DNA fragments are cloned into TA cloning vector (e.g., pGEM-T Easy), followed by transforming *Escherichia coli* DH5α competent cells. A known volume of transformed cells is spread plated on agar plate containing appropriate antibiotics. The insert from positive colonies is re-amplified and sequenced by standard Sanger's dideoxy method.

In diversity assessment, the major limitation of a typical clone library of 16S rDNA is analysis and sequencing of individual clones. A typical clone library of 16S rDNA contains nearly 1,000 clones that constitute only small portion of the overall diversity of an environmental sample. Previous study by Dunbar et al. (2002) has shown that an environmental sample, viz., soil necessitates the analysis of more than 40,000 clones to document 50% of total richness.

Denaturing or temperature-gradient gel electrophoresis (DGGE/TGGE)

DGGE is a molecular fingerprinting technique employed for determination of complex microbial diversity. This technique distinguishes short fragment of DNA based on their melting characteristic and applied to ascertain sequence variations in number of genes from various organisms. It relies on the fact that during electrophoresis, single stranded DNA migrates slower than double stranded similitude due to elevated interaction of exposed nucleotide to the gel matrix in the single stranded molecule. The procedure includes the electrophoretic migration of bacterial 16S rDNA fragments in polyacrylamide gel containing a linear increasing gradient of chemical denaturants (usually urea and formamide). As double stranded DNA molecule passes through the gradient of denaturants, at a particular denaturant concentration, a transition from helical to partially melted molecules occur (DNA molecules start to denature) and thereby their migration is retarded due to newly acquired branched structure (Fig. 4a). The differences in melting behavior cause the DNA fragments to stop migrating at different position in the denaturing gel and thereby effective separation. The denaturation depends on %GC content and therefore in this technique, DNA fragments that are identical in length but different in sequences can be separated.

DNA molecules containing low GC content may totally separate into single stranded (complete denaturation) and may not form any detectable band. Therefore, in a later modified method (Sheffield et al. 1989), to increase the detection sensitivity of single-base variation by DGGE, a 40-base-pair G+C-rich sequence (designated as GC-clamp) was incorporated by PCR onto the 5'-end of one of the primers. This results in partially melted structure in which GC-clamp prevents the complete denaturation (Fig. 4b).

This technique became very popular and has been successfully applied for microbial ecological studies such as profiling community complexity (Muyzer et al. 1993, Viszwapriya et al. 2015), to compare DNA extraction methods (Ariefdjohan et al. 2010, Dilhari et al. 2017), to study seasonal variation (Alonso-Sáez et al. 2007, Oberbeckmann et al. 2014), biofouling diversity (Ivnitsky et al. 2007, Belgini et al. 2018, Rajeev et al. 2019) and to assess the impacts of anthropogenic activities on bacterioplankton population (Jeffries et al. 2016).

One of the major limitations of this method is the separation of relatively short (500 bp) DNA fragments and therefore it is most commonly applied to analyze one or two of nine hyper-variable regions (V_1–V_9) of 16S rRNA gene, unlike other fingerprinting techniques that target the analysis of full-length 16S rRNA gene. PCR primers targeting the hypervariable region (V_3, V_3–V_4, V_6, V_6–V_8 and V_1–V_3) are

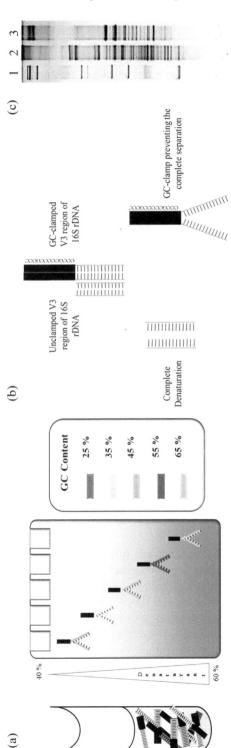

Fig. 4. Illustration of denaturing gradient gel electrophoresis: (a) Parallel DGGE gel showing increasing gradient (top-bottom) of denaturant (urea and formamide). Heterogeneous population of amplified hypervariable-16S rDNA fragments electrophoresed parallel to the direction of electrophoresis. Once the DNA molecule reaches at particular concentration of denaturant in the gel, molecules start to denature and their migration halt. High content of GC needs higher concentration of denaturant to achieve denaturation. (b) Generally, molecules containing lower GC content may not form any band due to complete denaturation; to prevent it, a GC-rich (typically 40 base pair long) DNA is added to the 5'-end of PCR amplicons. (c) DGGE employed to characterize and compare the planktonic (lane 2) and biofilm-forming (lane 3) bacterial diversity on artificial surfaces at the south coast of India (Rajeev et al. 2019). Metagenomic DNA extracted from both communities were used as templates for PCR amplification of hypervariable V3 region (~ 180 bp) of 16S rDNA and were applied onto 11% polyacrylamide gel. Each gel contained a linear gradient of denaturants urea and formamide (100% denaturant concentration corresponds to 7 M urea [w/v] and 40% [v/v] deionised formamide). Electrophoresis was performed at constant voltage (100 V) and temperature (60°C) for 15 h in 1X TAE buffer using INGENY PHORU system (The Netherlands). After electrophoresis, the gel was stained for 45 min in ethidium bromide (0.5 μg ml⁻¹), rinsed for 30 min in 1X TAE buffer and DGGE profile was visualized using gel documentation system.

used to amplify the intervening fragment(s) of 16S rDNA from uncharacterized bacterial community. The heterogeneous mixture of PCR amplified genes (mixed PCR products) is then separated by parallel DGGE. This allows the identification, comparison and relative abundance of dominant taxa present in any environmental sample (Fig. 4c).

In DGGE, each band represents a single microbial species (phylotype) and its intensity corresponds to the abundance. However, methodological limitations of DGGE should be noted. For example, DGGE fingerprinting relies on PCR amplification of 16S rDNA, which itself can be present in multiple copies with variation in sequences (Nubel et al. 1996) and therefore represent multiple bands even from single bacterial species. Similarly, co-migration of bands containing different DNA sequences but similar melting behavior can underestimate the community composition (Casamayor et al. 2000). Temperature gradient gel electrophoresis (TGGE) is another variant and relies on the same principle of DGGE except that a temperature gradient is applied to separate the DNA molecules rather than chemical denaturant.

Most of the published reports have targeted the hyper-variable V_3 region (Muyzer et al. 1993, Viszwapriya et al. 2015) of the 16S rRNA gene, as it is a relatively short fragment containing higher nucleotide diversity. The major advantages of these techniques are their affordability for ordinary laboratories, fast and the relative easy interpretation of the results. Moreover, individual bands of interest can be excised from the gel and representative phylotype can be identified through sequencing.

Terminal Restriction Fragment Length Polymorphism (T-RFLP)

T-RFLP is another PCR based molecular method for rapid analysis and comparison of complex microbial population in which an existing technique (amplified ribosomal DNA restriction analysis; ARDRA) is extended (Liu et al. 1997). In this method, the 16S rRNA gene of mixed bacterial population is PCR-amplified using fluorescent-labeled one or both of the primers. PCR products labeled at 5'-end are then digested with restriction endonuclease (enzymes that have 4 base pair recognition site). The size and relative abundance of the fluorescent-labeled terminal restriction fragments (T-RFs) are determined by capillary electrophoresis in an automated DNA sequencer (Fig. 5).

The result of this method is an electropherogram (graph representing the intensity plot) of capillary electrophoresis in which axis-x and axis-y represent the fragment's size and its intensity, respectively (Clement et al. 1998). Since differences in T-RFs depend on the sequence and length of the 16S rDNA, phylogenetic relation among the bacterial community can be resolved. A set of T-RFs with different lengths obtained from any environmental sample is referred to as T-RFs profile and considered as DNA fingerprinting of dominant microbial population present in that environmental sample.

Though the technique is comparatively accurate and reproducible for rapid characterization of unknown microbial community, it is hampered with few limitations that can prevent it from deciphering the accurate microbial community structure. Primarily, discrimination of bacterial population through this method relies on employed restriction enzyme(s). Typically, due to their higher frequency,

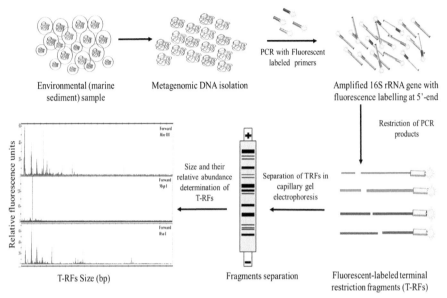

Environmental (marine sediment) sample

Metagenomic DNA isolation

PCR with Fluorescent labeled primers

Amplified 16S rRNA gene with fluorescence labelling at 5'-end

Restriction of PCR products

Size and their relative abundance determination of T-RFs

Separation of TRFs in capillary gel electrophoresis

Relative fluorescence units

Forward *Hae* III

Forward *Msp* I

Forward *Rsa* I

T-RFs Size (bp)

Fragments separation

Fluorescent-labeled terminal restriction fragments (T-RFs)

Fig. 5. Schematic representation of principle and workflow of terminal-restriction fragment length polymorphism (T-RFLP): Extracted metagenomic DNA from three marine sediment samples are taken for PCR amplification of marker 16S rRNA gene. During PCR, one (forward) of either primers that target the gene are fluorescently labeled at 5'-end with carboxyfluorescein (6-Fam). The fluorescent-labelled fragments subsequently digested with three restriction enzymes (*Hae* III, *Msp* I and *Rsa* I) and the resulting fragments are mixed with standard size markers. The length of the terminal-restriction fragments (T-RFs) was determined on an ABI PRISM 310 Genetic Analyzer (Applied Biosystems) (Data from Nithya and Pandian. 2012. Evaluation of bacterial diversity in Palk Bay sediments using terminal-restriction fragment length polymorphisms (T-RFLP). Applied Biochemistry and Biotechnology 167(6): 1763–1777).

restriction endonucleases that have four base pair recognition sites, viz., *Bst*U1, *Dde*I, *Sau*3A1 and *Msp*I are frequently used. It has been recommended that the use of more than one restriction enzymes can increase the detection limit (Schutte et al. 2008). Moreover, to evaluate the ability of any endonuclease enzyme to discriminate microbial assemblages, *in silico* digestion can be done using bioinformatics tools such as T-RFLP analysis programme (TAP) T-RFLP and MiCA for 16S rRNA gene. The TAP T-RFLP is located on the Ribosomal Database Project (RDP) website and facilitates the selection of the best enzymes(s)-primer combination.

Secondly, determination of accurate T-RFs is highly essential, especially when the goal of the investigation is to determine the plausible community composition. Generally, community composition can be determined by web-based tool, in which T-RFs sizes obtained experimentally from microbial community are compared with T-RFs generated from *in silico* digestion of 16S rDNA using same primers and enzymes. Another major limitation is the disparity between obtained and real length of T-RFs (Kaplan and Kitts 2003). Different fluorophores might affect the electrophoretic mobility of fragments in acrylamide matrix, raising the difficulties in accurate size determination. Few studies (Tu et al. 1998, Pandey et al. 2007) have reported the variations in generation of T-RFs length with the use of different fluorescent dyes used for labelling the T-RFLP primers.

In ARDRA, all the amplified 16S rDNA digested products resulting in unique band pattern are taken for community characterization, whereas, in T-RFLP, as the mixture of amplicons are sized by sequencer, only the terminal fragments labelled with fluorescent dye are read by laser to produce electropherogram while all other fragments are ignored and therefore the use of descriptor "terminal" in the name of technique.

Because of its simplicity, this method has been widely applied to examine the diversity of bacterial 16S rDNA (Katsivela et al. 2005, Hullar et al. 2006, Nithya and Pandian 2012), archaeal 16S rDNA (Kotsyurbenko et al. 2004, Lu et al. 2005), fungal ribosomal DNA (Johnson et al. 2004) and functional genes such as genes that code for methane oxidation and nitrogen fixation (Horz et al. 2000, Tan et al. 2003). However, most frequently, the technique utilizes the small subunit (16S or 18S) rRNA genes from mixed microbial population using PCR. The routinely used fluorescent dye that used to label the primers are 6-Carboxyfluorescein (6-FAM) and Hexachloro-fluorescein (HEX).

Automated ribosomal intergenic spacer analysis (ARISA)

A semi-automated genetic fingerprinting technique relies on Intergenic spacer (IGS) region of ribosomal RNA operon, located between small (16S) and large (23S) subunit (Fisher and Triplett 1999). Similar to 16S rRNA gene, it is used as bacterial barcode to study diversity analysis wherein IGS region of rRNA operon is employed. IGS fragment size ranges from 400–1200 bp and display more heterogeneity than flanked regions. ARISA facilitates differentiating bacterial strains and closely related species with the assistance of heterogeneity in both length and base arrangement of IGS region.

Bacterial rRNA operon of rRNA genetic locus consists of genes for 16S, 23S and 5S. Pre-rRNA, a transcript product of rRNA operon, has the sequence of 16S, spacer, tRNA, spacer, 23S, spacer and 5S gene ordering from 5'–3'. The IGS (also called internal transcribed spacer; ITS) exhibits large degree of variations (spacer polymorphisms) among closely related bacterial species. The spacer region also contains some genes of tRNA, that plays a major role in providing heterogeneity to the operon.

This method is an advancement of Ribosomal Intergenic Spacer Analysis (RISA) that differs in detection process. RISA has major drawbacks as it completely relies on gel-based comparison. To overcome the associated drawbacks with RISA, ARISA utilizes fluorescence labelled (e.g., phosphoramidite dye 5-FAM) primer that can be detected in an automated detector.

The following steps are followed in this technique (1) PCR amplification of the conserved regions present between 16S and 23S genes (flanking ITS-1) using fluorescently tagged universal oligonucleotide primers and metagenomic DNA as template (2) Electrophoretic separation of fluorescently labeled PCR products and subsequent detection on an automated sequencing system. The outcome of an ARISA is an electropherogram—a plot with peak height (fluorescence unit) vs. fragment length (Fig. 6). The relative quantity of each PCR product that is cloned and sequenced is determined by the ratio between fluorescence/peak area of the DNA

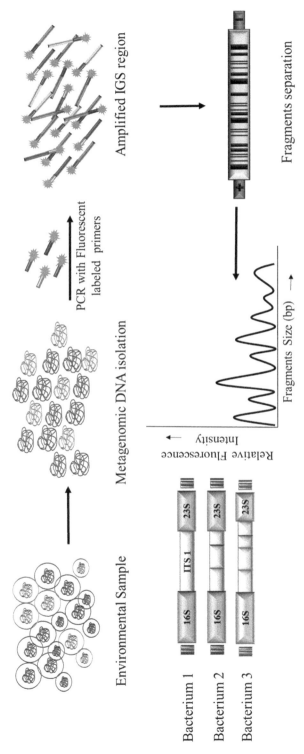

Fig. 6. Schematic representation of automated ribosomal intergenic spacer analysis (ARISA): A culture-independent technique involving isolation of metagenomic DNA and amplification of internal transcribed spacer (ITS) region using fluorescently labelled primers. The principle lies on heterogeneity of ITS-1 region among the closely related bacteria. Amplicons are run through capillary electrophoresis and the outcome is an electropherogram (a plot with peak height (fluorescence unit) vs. fragment length), which determines the bacterial diversity. The bands can be excised and sequenced for bacterial identification.

fragment to the total fluorescence of all the DNA fragments of the obtained bacterial community profile.

After 16S rRNA gene, ITS-1 region is used as major barcode for bacterial identification. ARISA technique established a great deal of attention in strain typing and bacterial diversity assessment owing to spacer polymorphisms in rRNA operon. This technique has been popularized in distinguishing bacteria at strain level in many studies that deals with assessment of ruminal bacterial diversity (Saro et al. 2018), dry biofilm bacterial diversity (Ledwoch et al. 2018), rhizosphere soil diversity (Mapelli et al. 2018), freshwater bacterial diversity (Fisher and Triplett 1999), bacterial community in catfish pond (Arias et al. 2006), diversity in raw milk and curd (Feligini et al. 2014), bacterial community structure in an aerated lagoon (Yu and Mohn 2001) and maple sap bacterial composition using multiplex ARISA (Filteau et al. 2011).

However, reports are available that discuss the potential issues inherent in this method. Major conflicting biases are (i) presence of multiple operons in single genome may differ in the length of the spacer, thus resulting in overestimation of diversity (ii) diversity may underestimate if different bacteria possess spacer region of identical length (iii) each primer pair targets its specific bacterial community composition (Kovacs et al. 2010).

Amplified ribosomal DNA restriction analysis (ARDRA)

ARDRA is an extended version of Restriction Fragment Length Polymorphisms (RFLP) that hires 16S rRNA gene or IGS region of rRNA operon for bacterial diversity and community composition analysis (Laguerre et al. 1994). Generally, the PCR products obtained from environmental samples are digested and resulting fragments are resolved on agarose or polyacrylamide gel (Smit et al. 1997). Although this method gives no idea about the type of microorganism present in any sample, it is still adapted for prompt monitoring of the shift of microbial community over time and scales.

This analysis tool determines the bacterial community and diversity with the aid of banding pattern of 16S rRNA gene that is digested with restriction endonucleases, especially tetra-base cutters (e.g., Alu I, Hae III, Mse l, Hinf 1, Sau 3A1, Hpal l). Tetra-base cutters are widely used as the occurrence of restriction site is more frequent and random (at every 256th base) compared to six (at every 4096 bp) and eight (at every 65536 bp) base cutters, which will be irrelevant with standard size of 16S rDNA (1500 bp).

Apart from this, it is a powerful and sensitive technique to classify and group a large collection of culturable bacterial strains into their respective phylotypes. This technique involves: (1) cultivation of bacterial targets *in vitro,* (2) DNA isolation of selected isolates, (3) amplification of 16S rRNA gene through PCR, (4) restriction of amplified products with suitable restriction enzyme(s) and (5) visualization of DNA banding polymorphisms through electrophoresis in high percent agarose gel or in polyacrylamide gel (Fig. 7).

This method is mainly limited only to culture-dependent studies with the fact that ARDRA profiling of mixed bacterial community species gives a complex-banding pattern as single species may give more than one band. This technique is

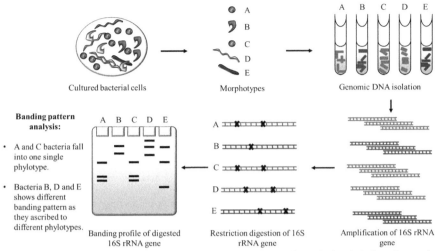

Fig. 7. Workflow of amplified ribosomal DNA restriction analysis: Figure depicts isolation of genomic DNA from five bacterial strains amplified with specific primer targeting 16S rRNA gene cluster. The amplicons are digested with 4 bp cutter restriction enzyme to reflect the difference in the restriction sites (vary with every individual bacteria). The digested fragments are electrophoresed to differentiate similar and dissimilar bacteria. The gel image states that the bacteria A and B are clustered into single phylotype (same banding pattern), while others (C, D, E) are different.

much useful in identification of bacterial collection that falls into same phylotype, ultimately reducing the chances of redundancy (re-sequencing the same bacterium). The bias may include same banding pattern of different but closely related bacterial strains and use of more than two restriction enzyme is recommended to overcome such biases.

ARDRA has been a potential DNA fingerprinting technique and is preferred for comparing diversity of the cultivable bacteria and its community composition in changing environment. Recent research using ARDRA comprises differentiation among Lactobacillus, Pediococcus and Weissella species (Adesulu-Dahunsi et al. 2017), genotypic and serotypic confirmations of bacterial community (Rajput et al. 2017), discriminating potentially probiotic Lactobacillus species from traditional dairy products (Ghafouri-Fard et al. 2018), soil characterization by bacterial community analysis for forensic applications (Habtom et al. 2017) and in combination with various biochemical tests for effective identification of bacterial species.

Single Stranded Conformation Polymorphism (SSCP)

SSCP is a screening method to identify different genomic variants, which was originally described by Orita et al. (1989) for the detection of point mutations. The procedure includes pre-labelling the sequence of interest and amplified simultaneously through PCR. Amplified products are denatured and electrophoretically separated on non-denaturant polyacrylamide gel (Fig. 8). SSCP relies on the secondary structure conformation of single stranded DNA that directly depends on sequence and its intramolecular interactions. DNA electrophoresis depends not only on the size and

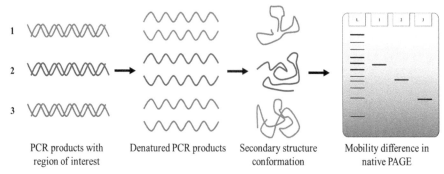

Fig. 8. Single stranded conformational polymorphism (SSCP): Schematic representation of the major steps involved in SSCP. The targeted region of the DNA is amplified and denatured by chemicals or with higher temperature. The change in the secondary structure of ssDNA with respect to its order of base pairs is unique to every bacterial taxa. They reflect the position of the band in a native polyacrylamide gel, which is displayed in the picture. L—denotes the ladder and 1, 2, 3 are the bacterial strains.

on molecular weight of the strand, but also on its shape. Even with same molecular weight, change in single nucleotide of DNA strand gives a different secondary conformation and thereby different position in the gel. This technique is often used in mutation detection (insertions and deletions). Lee et al. (1996) first described this method in bacterial diversity and community structure analysis.

Mobility variations of DNA strands due to its confirmation difference in the gel can lead to bacterial structure and diversity assessment of a particular habitat. Similar to DGGE/TGGE, each band from the gel can be eluted and sequenced for bacterial identification. SSCP has been successful in many prokaryotic and eukaryotic DNA mutation identification and is efficacious in bacterial diversity appraisal.

This method has been frequently utilized in characterization of soil bacterial diversity (Smalla et al. 2007), colonic mucosa associated bacterial microflora (Ott et al. 2004), bacterial diversity in maize rhizospheres (Schmalenberger and Tebbe 2003), diversity of a natural iron-fertilized phytoplankton bloom in the Southern Ocean (West et al. 2008), diversity difference in finished compost and vermicompost (Fracchia et al. 2006) and bacterial community structure in hot water biofilm (Farhat et al. 2018).

SSCP does not require any GC-clamp or restriction making it a more feasible and easy technique in diversity analysis. Moreover, advances in SSCP by automating through ABI Prism capillary array sequencers has further facilitated its analysis to be much easier. Failings in SSCP encompasses prohibitive rate of annealing of DNA strands within fraction of seconds when the used concentration of DNA is in excess. Denatured strands give rise to two bands with same mobility or different, which may lead to misinterpretation of results. The bands in gel can be increased owing to different conformations of same strand or be decreased due to coexisting conformation of different strands. Moreover, the secondary structure mobility is a difficult task to predict as its mobility is defined not only by its primary structure, but also by variables such as temperature and ionic strength. Hence, these factors are more important for reproducible results.

Non-PCR based techniques

Techniques under non-PCR based approach involves the use of DNA probe(s) for bacterial identification. Molecular markers having specific sequence arrangements for each bacterial taxa assist in probe designing and act as principle behind these techniques in bacterial diversity analysis. Non-PCR based techniques include FISH and DNA microarray that depend on DNA probes (known oligonucleotide sequence) and primers for bacterial identification through different DNA hybridization procedures, thus avoiding PCR and its associated biases.

Fluorescence in situ hybridization (FISH)

FISH enables *in situ* identification and enumeration of individual microbial cells using oligonucleotide probes (Amann et al. 1995). This technique directly visualizes the defined sequences of DNA on morphologically preserved cell strains. The basic objective of FISH is to inspect the presence/absence of targeted DNA at cellular level by facilitating specific detection and enumeration of bacterial community directly in the environment that circumvents the need for cultivation. The principle behind FISH is the ability of single stranded fluorescently labelled DNA probe to hybridize effectively with the exact complementary DNA sequence of the cell.

Briefly, the protocol follows fixing the bacterial strains using fixatives (formaldehyde) followed by permeablizing the cells using lysozyme for the probe to penetrate and hybridize. After an optimized time required for hybridization, the cells are viewed under confocal laser scanning microscope (CLSM) or other fluorescence microscope (Fig. 9). The major stains used are: DAPI (4, 6-diamidino-2-phenylindole; emits blue color), propidium iodide/rhodamine (for red) and Fluorescein isothiocyanate (FITC) for green.

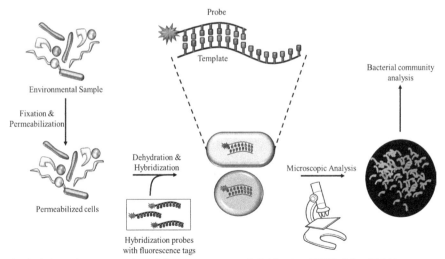

Fig. 9. Schematic representation of fluorescence *in situ* hybridization (FISH): Microbial biomass are fixed and permeabilized to access target DNA. A target specific probe is hybridized to target DNA and visualized under fluorescence microscope.

This technique highly relies on the probe and by virtue of the molecular marker-16S rRNA gene with conserved and variable regions. FISH probes are 18–30 nucleotides long oligonucleotides that contain a fluorescent dye at 5'-end and can be designed at any phylogenetic level targeting from domain to species-specific. FISH has been majorly applied in the study of bacterial biofilm to analyze the spatial pattern, especially in multispecies biofilm with multicolor FISH (mFISH) (Thurnheer et al. 2004, Almeida et al. 2011, de Paz 2012). Bacterial diversity from sponges (Webster et al. 2001, Li et al. 2006), human gut microbiota (Aminov et al. 2006, Manichanh et al. 2006), environmental samples (Daims et al. 2001, Pernthaler and Amann 2004) and various other fields has been analysed by employing FISH. The use of fluorescence in FISH overcomes the drawbacks of radiolabelling probes that was previously used in *in situ* hybridization techniques. FISH also overcomes the PCR bias but lacks specificity and fails in the detection of the least abundant bacterial community, as it needs a visible DNA for binding.

DNA microarray

Array based techniques monitor the interactions between molecules such as DNA-DNA, protein-protein, antibody-antibody and their combinatorial communications with the aid of prearranged and programmed library of molecular probes. The original DNA array was primarily shaped with the colony hybridization method of Grunstein and Hogness (1975). Later, Gergen et al. (1979) picked this technique and adapted an ordered array of colonies in 144-well microplates. DNA microarray/ DNA chip facilitates bacterial community analysis and identification through DNA-DNA hybridization with the aid of hundreds to thousands of unique DNA probes pre-arranged in a 2D/3D surface.

The PCR products obtained from environmental samples are allowed to hybridize with probes and are subjected to CLSM (Gentry et al. 2006). In this method, color signals and their intensity directly correspond to the number of bacterial species and their abundance, respectively.

Cross hybridization, particularly when dealing with environmental samples, is a major limitation of this technique. Further, bacterial taxa can completely be ignored if microarray does not have specific probe. Novel taxa cannot be detected.

DNA microarrays provide the output similar to latest and highly sophisticated next generation sequencing technology. DNA microarrays have been widely applied to study diversity and gene expression analysis. In the field of bacterial diversity, DNA microarray technology is majorly classified into two, based on the design of the probe: (i) 16S rRNA gene microarrays (Phylochip) and (ii) functional gene arrays. DNA microarray vastly exploits the diversity of bacteria in various fields of environmental studies. Rastogi et al. (2010) assessed the microbial diversity of mining impacted soils from uranium mine sites, where the extensive phylogenetic diversity in soil affected by uranium was revealed. Cruz-Martínez et al. (2009) exploited the seasonal changes in soil microbial consortia in grassland. Sagaram et al. (2009) employed phylochip arrays in bacterial diversity analysis of Huanglongbing pathogen-infected citrus, where the phylochip data revealed the abundant bacterial taxa in symptomatic midribs and in asymptomatic midribs of affected plant leaf. Corals were subjected to bacterial diversity assessment by Sunagawa et al. (2009) and Closek et al. (2014), where they divulged the bacterial community profiles responsible for causing Yellow Band Disease and White Plague Disease, respectively. Further, DNA microarray was also employed in assessment of diversity in wastewater treatment plants (Kelly et al. 2005), marine sediments (Wu et al. 2008) and human diseases (Paster et al. 2006).

Next-generation sequencing (NGS): a revolution in the era of microbial ecology

Though above-mentioned culture-independent techniques provide a comprehensive overview of microbial community structure in various environments, all methods are very laborious, erroneous and hampered with various other limitations such as identification of only dominant (abundant) microorganisms, not rare taxa. Advent of NGS technology has tremendously revolutionized the field of microbial ecology and broadened the scientific vision towards complex community composition. The term NGS refers to massively parallel sequencing technologies that include various commercialized platforms such as 454/Roche, Illumina/Solexa, Ion Torrent and ABI/SOLiD. In this section, the details of each NGS platforms are not described, although interested readers are directed to Chapter 9 by Sushmitha et al., where the sequencing chemistries and applications of NGS platforms in relation to bacterial diversity assessment are explained.

The main characteristics of these technologies, such as generating millions of reads (high throughput) at a relatively rapid rate and lesser costs, have greatly transformed the field of microbial ecology. In the field of microbial ecology, these high-throughput technologies are useful in various environmental studies that deal with metagenomics, metatranscriptomics, proteogenomics, amplicons sequencing and whole-genome shotgun sequencing.

Concluding remarks and future directions

Advent of molecular genomic tools has greatly revolutionized the field of microbial ecology. The frequency of studies carried out to study microbial diversity and their unique role in various ecosystems uncover that we have only explored the superficial surface of the huge genetic diversity present on the Earth. Several fundamental questions, viz.: "How many number of microbial species are there on the Earth?", "Which factors shape the community structure?", "To what extend their metabolic activities are diverse?" are still of great ecological concern. Assessment of microbial community composition in an environmental sample is a great challenge due to the chances of error at various levels such as extraction of metagenomic DNA from environmental sample, number of samples for analysis, influence of other environmental factors that are not under consideration, PCR amplification, technical issues and limited tools for result analysis and interpretation.

All of the molecular tools available for community structure analysis are hampered with limitations, and none of the method provides a comprehensive and overall picture of complex microbial community. However, it has been highly recommended that any investigation should deal with both culture-dependent and -independent methods, particularly in the studies that deal with effect of anthropogenic activities on microbial diversity.

Advancements in NGS has already surpassed the use of molecular finger-printing methods and significantly contributed towards the understating of microbial ecology. The launch of frugal innovative technology such as Oxford's Nanopore is expected to be a superior culture-independent approach in analysing microbial community. Integrated technologies of different meta-omic approaches, viz., metagenomics, metatranscriptomics and metaproteomics with *in situ* geochemistry could afford a more comprehensive data on host microbiome and indigenous microbial diversity along with their potential functions. Moreover, single cell genomics is reaching its peak with its ability to define microbial community at species level. Considering the fast leap of recent advancements, microbial diversity associated with any organism and/or habitat has become an attainable scientific goal.

Acknowledgements

The authors sincerely acknowledge the computational and bioinformatics facility provided by the Alagappa University Bioinformatics Infrastructure Facility (funded by DBT, GOI; File No. BT/BI/25/012/2012, BIF). The authors also thankfully acknowledge DST-FIST (Grant No. SR/FST/LSI-639/2015(C)), UGC-SAP (Grant No. F.5-1/2018/DRS-II (SAP-II)) and DST-PURSE (Grant No. SR/PURSE Phase 2/38 (G)) for providing instrumentation facilities. Financial support provided to MR by RUSA Phase 2.0 [F.24-51/2014-U, Policy (TN Multi-Gen), Dept. of Edn, GoI] in the form of Ph.D. Fellowship is also thankfully acknowledged.

References

Adesulu-Dahunsi, A. T., Sanni, A. I. and Jeyaram, K. 2017. Rapid differentiation among Lactobacillus, Pediococcus and Weissella species from some Nigerian indigenous fermented foods. LWT-Food Science and Technology 77: 39–44.

Almeida, C., Azevedo, N. F., Santos, S., Keevil, C. W. and Vieira, M. J. 2011. Discriminating multi-species populations in biofilms with peptide nucleic acid fluorescence *in situ* hybridization (PNA FISH). PLoS One 6(3): e14786.

Alonso-Sáez, L., Balagué, V., Sà, E. L., Sánchez, O., González, J. M., Pinhassi, J. et al. 2007. Seasonality in bacterial diversity in north-west Mediterranean coastal waters: assessment through clone libraries, fingerprinting and FISH. FEMS Microbiology Ecology 60(1): 98–112.

Amann, R. I., Ludwig, W. and Schleifer, K. H. 1995. Phylogenetic identification and *in situ* detection of individual microbial cells without cultivation. Microbiology and Molecular Biology Reviews 59(1): 143–169.

Aminov, R. I., Walker, A. W., Duncan, S. H., Harmsen, H. J., Welling, G. W. and Flint, H. J. 2006. Molecular diversity, cultivation, and improved detection by fluorescent *in situ* hybridization of a dominant group of human gut bacteria related to *Roseburia* spp. or Eubacterium rectale. Applied and Environmental Microbiology 72(9): 6371–6376.

Arias, C. R., Abernathy, J. W. and Liu, Z. 2006. Combined use of 16S ribosomal DNA and automated ribosomal intergenic spacer analysis to study the bacterial community in catfish ponds. Letters in Applied Microbiology 43(3): 287–292.

Ariefdjohan, Merlin, W., Dennis A. Savaiano and Cindy H. Nakatsu. 2010. Comparison of DNA extraction kits for PCR-DGGE analysis of human intestinal microbial communities from fecal specimens. Nutrition Journal 9(1): 23.

Belgini, D. R., Siqueira, V. M., Oliveira, D. M., Fonseca, S. G., Piccin-Santos, V., Dias, R. S. et al. 2018. Integrated diversity analysis of the microbial community in a reverse osmosis system from a Brazilian oil refinery. Systematic and Applied Microbiology 41(5): 473–486.

Bochner, B. 1989. Sleuthing out bacterial identities. Nature 339(6220): 157–158.

Bossio, D. A., Scow, K. M., Gunapala, N. and Graham, K. J. 1998. Determinants of soil microbial communities: effects of agricultural management, season, and soil type on phospholipid fatty acid profiles. Microbial Ecology 36(1): 1–12.

Cai, L., Ye, L., Tong, A. H. Y., Lok, S. and Zhang, T. 2013. Biased diversity metrics revealed by bacterial 16S pyrotags derived from different primer sets. PloS ONE 8(1): e53649.

Casamayor, E. O., Schäfer, H., Bañeras, L., Pedrós-Alió, C. and Muyzer, G. 2000. Identification of and spatio-temporal differences between microbial assemblages from two neighboring sulfurous lakes: comparison by microscopy and denaturing gradient gel electrophoresis. Applied and Environmental Microbiology 66(2): 499–508.

Christian, B. W. and Lind, O. T. 2006. Key issues concerning Biolog use for aerobic and anaerobic freshwater bacterial community-level physiological profiling. International Review of Hydrobiology 91(3): 257–268.

Clement, Brian G., Lucia E. Kehl, Kristin L. DeBord and Christopher L. Kitts. 1998. Terminal restriction fragment patterns (TRFPs), a rapid, PCR-based method for the comparison of complex bacterial communities.Journal of Microbiological Methods 31(3): 135–142.

Closek, C. J., Sunagawa, S., DeSalvo, M. K., Piceno, Y. M., DeSantis, T. Z., Brodie, E. L. et al. 2014. Coral transcriptome and bacterial community profiles reveal distinct Yellow Band Disease states in Orbicella faveolata. The ISME Journal 8(12): 2411–2422.

Cruz-Martínez, K., Suttle, K. B., Brodie, E. L., Power, M. E., Andersen, G. L. and Banfield, J. F. 2009. Despite strong seasonal responses, soil microbial consortia are more resilient to long-term changes in rainfall than overlying grassland. The ISME Journal 3(6): 738–744.

Daims, H., Ramsing, N. B., Schleifer, K. H. and Wagner, M. 2001. Cultivation-independent, semiautomatic determination of absolute bacterial cell numbers in environmental samples by fluorescence *in situ* hybridization. Applied and Environmental Microbiology 67(12): 5810–5818.

de Paz, L. E. C. 2012. Development of a multispecies biofilm community by four root canal bacteria. Journal of Endodontics 38(3): 318–323.

Derry, A. M., Staddon, W. J. and Trevors, J. T. 1998. Functional diversity and community structure of microorganisms in uncontaminated and creosote-contaminated soils as determined by sole-carbon-source-utilization. World Journal of Microbiology and Biotechnology 14(4): 571–578.

Derry, A. M., Staddon, W. J., Kevan, P. G. and Trevors, J. T. 1999. Functional diversity and community structure of microorganisms in three arctic soils as determined by sole-carbon-source-utilization. Biodiversity & Conservation 8(2): 205–221.

DeSantis, T. Z., Brodie, E. L., Moberg, J. P., Zubieta, I. X., Piceno, Y. M. and Andersen, G. L. 2007. High-density universal 16S rRNA microarray analysis reveals broader diversity than typical clone library when sampling the environment. Microbial Ecology 53(3): 371–383.

Dilhari, A., Sampath, A., Gunasekara, C., Fernando, N., Weerasekara, D., Sissons, C. et al. 2017. Evaluation of the impact of six different DNA extraction methods for the representation of the microbial community associated with human chronic wound infections using a gel-based DNA profiling method. AMB Express 7(1): 179.

Ellis, R. J., Thompson, I. P. and Bailey, M. J. 1995. Metabolic profiling as a means of characterizing plant-associated microbial communities. FEMS Microbiology Ecology 16(1): 9–17.

Farhat, M., Moletta-Denat, M., Trouilhé, M. C., Frère, J. and Robine, E. 2018. Transitory change of bacterial community structure in hot water biofilm: Effects of anti-legionella treatments. CLEAN–Soil, Air, Water 46(6): 1700203.

Feligini, M., Brambati, E., Panelli, S., Ghitti, M., Sacchi, R., Capelli, E. et al. 2014. One-year investigation of Clostridium spp. occurrence in raw milk and curd of Grana Padano cheese by the automated ribosomal intergenic spacer analysis. Food Control 42: 71–77.

Filteau, M., Lagacé, L., LaPointe, G. and Roy, D. 2011. Correlation of maple sap composition with bacterial and fungal communities determined by multiplex automated ribosomal intergenic spacer analysis (MARISA). Food Microbiology 28(5): 980–989.

Fisher, M. M. and Triplett, E. W. 1999. Automated approach for ribosomal intergenic spacer analysis of microbial diversity and its application to freshwater bacterial communities. Applied and Environmental Microbiology 65(10): 4630–4636.

Fracchia, L., Dohrmann, A. B., Martinotti, M. G. and Tebbe, C. C. 2006. Bacterial diversity in a finished compost and vermicompost: differences revealed by cultivation-independent analyses of PCR-amplified 16S rRNA genes. Applied Microbiology and Biotechnology 71(6): 942–952.

Frostegård, Å., Tunlid, A. and Bååth, E. 1993. Phospholipid fatty acid composition, biomass, and activity of microbial communities from two soil types experimentally exposed to different heavy metals. Applied and Environmental Microbiology 59(11): 3605–3617.

Garland, J. L. and Mills, A. L. 1991. Classification and characterization of heterotrophic microbial communities on the basis of patterns of community-level sole-carbon-source utilization. Applied and Environmental Microbiology 57(8): 2351–2359.

Garland, J. L. 1996. Patterns of potential C source utilization by rhizosphere communities. Soil Biology and Biochemistry 28(2): 223–230.

Gentry, T. J., Wickham, G. S., Schadt, C. W., He, Z. and Zhou, J. 2006. Microarray applications in microbial ecology research. Microbial Ecology 52(2): 159–175.

Gergen, J. P., Stern, R. H. and Wensink, P. C. 1979. Filter replicas and permanent collections of recombinant DNA plasmids. Nucleic Acids Research 7(8): 2115–2136.

Ghafouri-Fard, S., Hejazi, M. A., Afshar, D., Barzegari, A. and Eslami, S. 2018. 16S-amplified ribosomal DNA restriction analysis assay for discriminating potentially probiotic Lactobacillus species isolated from traditional dairy products. Journal of Kerman University of Medical Sciences 25(2): 175–186.

Gilbert, J. A., Jansson, J. K. and Knight, R. 2014. The Earth Microbiome project: successes and aspirations. BMC Biology 12(1): 69.

Grunstein, M. and Hogness, D. S. 1975. Colony hybridization: a method for the isolation of cloned DNAs that contain a specific gene. Proceedings of the National Academy of Sciences 72(10): 3961–3965.

Haack, S. K., Garchow, H., Klug, M. J. and Forney, L. J. 1995. Analysis of factors affecting the accuracy, reproducibility, and interpretation of microbial community carbon source utilization patterns. Applied and Environmental Microbiology 61(4): 1458–1468.

Habtom, H., Demanèche, S., Dawson, L., Azulay, C., Matan, O., Robe, P. et al. 2017. Soil characterisation by bacterial community analysis for forensic applications: A quantitative comparison of environmental technologies. Forensic Science International: Genetics 26: 21–29.

Horz, H. P., Rotthauwe, J. H., Lukow, T. and Liesack, W. 2000. Identification of major subgroups of ammonia-oxidizing bacteria in environmental samples by T-RFLP analysis of amoA PCR products. Journal of Microbiological Methods 39(3): 197–204.

Hullar, M. A., Kaplan, L. A. and Stahl, D. A. 2006. Recurring seasonal dynamics of microbial communities in stream habitats. Applied and Environmental Microbiology 72(1): 713–722.

Ivnitsky, H., Katz, I., Minz, D., Volvovic, G., Shimoni, E., Kesselman, E. et al. 2007. Bacterial community composition and structure of biofilms developing on nanofiltration membranes applied to wastewater treatment. Water Research 41(17): 3924–3935.

Jeffries, T. C., Schmitz Fontes, M. L., Harrison, D. P., Van-Dongen-Vogels, V., Eyre, B. D., Ralph, P. J. et al. 2016. Bacterioplankton dynamics within a large anthropogenically impacted urban estuary. Frontiers in Microbiology 6: 1438.

Johnson, D., Vandenkoornhuyse, P. J., Leake, J. R., Gilbert, L., Booth, R. E., Grime J. P. et al. 2004. Plant communities affect arbuscular mycorrhizal fungal diversity and community composition in grassland microcosms. New Phytologist 161(2): 503–515.

Kaplan, Christopher W. and Christopher L. Kitts. 2003. Variation between observed and true terminal restriction fragment length is dependent on true TRF length and purine content. Journal of Microbiological Methods 54(1): 121–125.

Katsivela, E., Moore, E. R. B., Maroukli, D., Strömpl, C., Pieper, D. and Kalogerakis, N. 2005. Bacterial community dynamics during *in-situ* bioremediation of petroleum waste sludge in land farming sites. Biodegradation 16(2): 169–180.

Kelly, J. J., Siripong, S., McCormack, J., Janus, L. R., Urakawa, H., El Fantroussi, S. et al. 2005. DNA microarray detection of nitrifying bacterial 16S rRNA in wastewater treatment plant samples. Water Research 39(14): 3229–3238.

Kotsyurbenko, O. R., Chin, K. J., Glagolev, M. V., Stubner, S., Simankova, M. V., Nozhevnikova, A. N. et al. 2004. Acetoclastic and hydrogenotrophic methane production and methanogenic populations in an acidic West-Siberian peat bog. Environmental Microbiology 6(11): 1159–1173.

Kovacs, A., Yacoby, K. and Gophna, U. 2010. A systematic assessment of automated ribosomal intergenic spacer analysis (ARISA) as a tool for estimating bacterial richness. Research in Microbiology 161(3): 192–197.

Laguerre, G., Allard, M. R., Revoy, F. and Amarger, N. 1994. Rapid identification of rhizobia by restriction fragment length polymorphism analysis of PCR-amplified 16S rRNA genes. Applied and Environmental Microbiology 60(1): 56–63.

Ledwoch, K., Dancer, S. J., Otter, J. A., Kerr, K., Roposte, D. and Rushton, L. 2018. Beware biofilm! Dry biofilms containing bacterial pathogens on multiple healthcare surfaces; a multi-centre study. Journal of Hospital Infection 100(3): 47–56.

Lee, D. H., Zo, Y. G. and Kim, S. J. 1996. Nonradioactive method to study genetic profiles of natural bacterial communities by PCR-single-strand-conformation polymorphism. Applied and Environmental Microbiology 62(9): 3112–3120.

Li, H., Zhang, Y., Li, D. S., Xu, H., Chen, G. X. and Zhang, C. G. 2009. Comparisons of different hypervariable regions of rrs genes for fingerprinting of microbial communities in paddy soils. Soil Biology and Biochemistry 41(5): 954–968.

Li, Z. Y., He, L. M., Wu, J. and Jiang, Q. 2006. Bacterial community diversity associated with four marine sponges from the South China Sea based on 16S rDNA-DGGE fingerprinting. Journal of Experimental Marine Biology and Ecology 329(1): 75–85.

Liu, W. T., Marsh, T. L., Cheng, H. and Forney, L. J. 1997. Characterization of microbial diversity by determining terminal restriction fragment length polymorphisms of genes encoding 16S rRNA. Applied and Environmental Microbiology 63(11): 4516–4522.

Lladó, S. and Baldrian, P. 2017. Community-level physiological profiling analyses show potential to identify the copiotrophic bacteria present in soil environments. PLoS ONE 12(2): e0171638.

Loreau, M., Naeem, S., Inchausti, P., Bengtsson, J., Grime, J. P., Hector, A. et al. 2001. Biodiversity and ecosystem functioning: current knowledge and future challenges. Science 294(5543): 804–808.

Lu, Y., Lueders, T., Friedrich, M. W. and Conrad, R. 2005. Detecting active methanogenic populations on rice roots using stable isotope probing. Environmental Microbiology 7(3): 326–336.

Manichanh, C., Rigottier-Gois, L., Bonnaud, E., Gloux, K., Pelletier, E., Frangeul, L. et al. 2006. Reduced diversity of faecal microbiota in Crohn's disease revealed by a metagenomic approach. Gut 55(2): 205–211.

Mapelli, F., Marasco, R., Fusi, M., Scaglia, B., Tsiamis, G., Rolli, E. and Erlich, H. 2018. The stage of soil development modulates rhizosphere effect along a High Arctic desert chronosequence. The ISME Journal 12(5): 1188–1198.

Mondini, C. and Insam, H. 2003. Community level physiological profiling as a tool to evaluate compost maturity: a kinetic approach. European Journal of Soil Biology 39(3): 141–148.

Mullis, K., Faloona, F., Scharf, S., Saiki, R. K., Horn, G. T. et al. 1986. Specific enzymatic amplification of DNA *in vitro*: the polymerase chain reaction. In Cold Spring Harbor Symposia on Quantitative Biology 51: 263–273. Cold Spring Harbor Laboratory Press.

Muyzer, G., De Waal, E. C. and Uitterlinden, A. G. 1993. Profiling of complex microbial populations by denaturing gradient gel electrophoresis analysis of polymerase chain reaction-amplified genes coding for 16S rRNA. Applied and Environmental Microbiology 59(3): 695–700.

Nguyen, N. P., Warnow, T., Pop, M. and White, B. 2016. A perspective on 16S rRNA operational taxonomic unit clustering using sequence similarity. NPJ Biofilms and Microbiomes 2(1): 1–8.

Nithya, C. and Pandian, S. K. 2012. Evaluation of bacterial diversity in Palk Bay sediments using terminal-restriction fragment length polymorphisms (T-RFLP). Applied Biochemistry and Biotechnology 167(6): 1763–1777.

Nocker, A., Burr, M. and Camper, A. K. 2007. Genotypic microbial community profiling a critical technical review. Microbial Ecology 54(2): 276–289.

Nubel, U., Engelen, B., Felske, A., Snaidr, J., Wieshuber, A., Amann, R. I. et al. 1996. Sequence heterogeneities of genes encoding 16S rRNAs in Paenibacillus polymyxa detected by temperature gradient gel electrophoresis. Journal of Bacteriology 178(19): 5636–5643.

Oberbeckmann, S., Loeder, M. G., Gerdts, G. and Osborn, A. M. 2014. Spatial and seasonal variation in diversity and structure of microbial biofilms on marine plastics in Northern European waters. FEMS Microbiology Ecology 90(2): 478–492.

Orita, M., Suzuki, Y., Sekiya, T. and Hayashi, K. 1989. Rapid and sensitive detection of point mutations and DNA polymorphisms using the polymerase chain reaction. Genomics 5(4): 874–879.

Ott, S. J., Musfeldt, M., Wenderoth, D. F., Hampe, J., Brant, O., Fölsch, U. R. et al. 2004. Reduction in diversity of the colonic mucosa associated bacterial microflora in patients with active inflammatory bowel disease. Gut 53(5): 685–693.

Pandey, J., Ganesan, K. and Jain, R. K. 2007. Variations in T-RFLP profiles with differing chemistries of fluorescent dyes used for labeling the PCR primers. Journal of Microbiological Methods 68(3): 633–638.

Paster, B. J., Olsen, I., Aas, J. A. and Dewhirst, F. E. 2006. The breadth of bacterial diversity in the human periodontal pocket and other oral sites. Periodontology 42(1): 80–87.

Pernthaler, A. and Amann, R. 2004. Simultaneous fluorescence *in situ* hybridization of mRNA and rRNA in environmental bacteria. Applied and Environmental Microbiology 70(9): 5426–5433.

Rajeev, M., Sushmitha, T. J., Toleti, S. R. and Pandian, S. K. 2019. Culture dependent and independent analysis and appraisal of early stage biofilm-forming bacterial community composition in the Southern coastal seawater of India. Science of the Total Environment 666: 308–320.

Rajput, Y., Neral, A. and Rai, V. 2015. Genotypic and serotypic confirmations of bacterial community to Kotumsar cave for occupational safety of cave workers and visitors from pathogenic threats. International Journal of Occupational Safety and Health 5(1): 22–27.

Rastogi, G., Osman, S., Vaishampayan, P. A., Andersen, G. L., Stetler, L. D. and Sani, R. K. 2010. Microbial diversity in uranium mining-impacted soils as revealed by high-density 16S microarray and clone library. Microbial Ecology 59(1): 94–108.

Reichardt, W., Mascarina, G., Padre, B. and Doll, J. 1997. Microbial communities of continuously cropped, irrigated rice fields. Applied and Environmental Microbiology 63(1): 233–238.

Sagaram, U. S., DeAngelis, K. M., Trivedi, P., Andersen, G. L., Lu, S. E. and Wang, N. 2009. Bacterial diversity analysis of Huanglongbing pathogen-infected citrus, using PhyloChip arrays and 16S rRNA gene clone library sequencing. Applied and Environmental Microbiology 75(6): 1566–1574.

Šajbidor, J. 1997. Effect of some environmental factors on the content and composition of microbial membrane lipids. Critical Reviews in Biotechnology 17(2): 87–103.

Saro, C., Molina-Alcaide, E., Abecia, L., Ranilla, M. J. and Carro, M. D. 2018. Comparison of automated ribosomal intergenic spacer analysis (ARISA) and denaturing gradient gel electrophoresis (DGGE) techniques for analysing the influence of diet on ruminal bacterial diversity. Archives of Animal Nutrition 72(2): 85–99.

Schmalenberger, A. and Tebbe, C. C. 2003. Bacterial diversity in maize rhizospheres: conclusions on the use of genetic profiles based on PCR-amplified partial small subunit rRNA genes in ecological studies. Molecular Ecology 12(1): 251–262.

Schütte, U. M., Abdo, Z., Bent, S. J., Shyu, C., Williams, C. J., Pierson, J. D. et al. 2008. Advances in the use of terminal restriction fragment length polymorphism (T-RFLP) analysis of 16S rRNA genes to characterize microbial communities. Applied and Environmental Microbiology 80(3): 365–380.

Sheffield, V. C., Cox, D. R., Lerman, L. S. and Myers, R. M. 1989. Attachment of a 40-base-pair G+ C-rich sequence (GC-clamp) to genomic DNA fragments by the polymerase chain reaction results in improved detection of single-base changes. Proceedings of the National Academy of Sciences 86(1): 232–236.

Singh, B. K., Campbell, C. D., Sorenson, S. J. and Zhou, J. 2009. Soil genomics. Nature Reviews Microbiology 7(10): 756.

Smalla, K., Oros-Sichler, M., Milling, A., Heuer, H., Baumgarte, S., Becker, R. et al. 2007. Bacterial diversity of soils assessed by DGGE, T-RFLP and SSCP fingerprints of PCR-amplified 16S rRNA gene fragments: do the different methods provide similar results? Journal of Microbiological Methods 69(3): 470–479.

Smit, E., Leeflang, P. and Wernars, K. 1997. Detection of shifts in microbial community structure and diversity in soil caused by copper contamination using amplified ribosomal DNA restriction analysis. FEMS Microbiology Ecology 23(3): 249–261.

Sogin, M. L., Morrison, H. G., Huber, J. A., Welch, D. M., Huse, S. M. and Neal, P. R. 2006. Microbial diversity in the deep sea and the underexplored "rare biosphere". Proceedings of the National Academy of Sciences 103(32): 12115–12120.

Staley, J. T. and Konopka, A. 1985. Measurement of *in situ* activities of nonphotosynthetic microorganisms in aquatic and terrestrial habitats. Annual Review of Microbiology 39(1): 321–346.

Sunagawa, S., DeSantis, T. Z., Piceno, Y. M., Brodie, E. L., DeSalvo, M. K., Voolstra, C. R. et al. 2009. Bacterial diversity and white plague disease-associated community changes in the Caribbean coral Montastraea faveolata. The ISME Journal 3(5): 512–521.

Tan, Z., Hurek, T. and Reinhold-Hurek, B. 2003. Effect of N-fertilization, plant genotype and environmental conditions on nifH gene pools in roots of rice. Environmental Microbiology 5(10): 1009–1015.

Thurnheer, T., Gmür, R. and Guggenheim, B. 2004. Multiplex FISH analysis of a six-species bacterial biofilm. Journal of Microbiological Methods 56(1): 37–47.

Tu, O., Knott, T., Marsh, M., Bechtol, K., Harris, D., Barker, D. et al. 1998. The influence of fluorescent dye structure on the electrophoretic mobility of end-labeled DNA. Nucleic Acids Research 26(11): 2797–2802.

Viszwapriya, D., Aravindraja, C. and Pandian, S. K. 2015. Comparative assessment of bacterial diversity associated with co-occurring eukaryotic hosts of Palk Bay origin. Indian Journal of Experimental Biology 53: 417–423.

Ward, D. M., Weller, R. and Bateson, M. M. 1990. 16S rRNA sequences reveal numerous uncultured microorganisms in a natural community. Nature 345(6270): 63–65.

Webster, N. S., Wilson, K. J., Blackall, L. L. and Hill, R. T. 2001. Phylogenetic diversity of bacteria associated with the marine sponge Rhopaloeides odorabile. Applied and Environmental Microbiology 67(1): 434–444.

West, N. J., Obernosterer, I., Zemb, O. and Lebaron, P. 2008. Major differences of bacterial diversity and activity inside and outside of a natural iron-fertilized phytoplankton bloom in the Southern Ocean. Environmental Microbiology 10(3): 738–756.

Whitman, W. B., Coleman, D. C. and Wiebe, W. J. 1998. Prokaryotes: the unseen majority. Proceedings of the National Academy of Sciences 95(12): 6578–6583.

Wu, L., Kellogg, L., Devol, A. H., Tiedje, J. M. and Zhou, J. 2008. Microarray-based characterization of microbial community functional structure and heterogeneity in marine sediments from the Gulf of Mexico. Applied and Environmental Microbiology 74(14): 4516–4529.

Yang, B., Wang, Y. and Qian, P. Y. 2016. Sensitivity and correlation of hypervariable regions in 16S rRNA genes in phylogenetic analysis. BMC Bioinformatics 17(1): 135.

Yu, Z. and Mohn, W. W. 2001. Bacterial diversity and community structure in an aerated lagoon revealed by ribosomal intergenic spacer analyses and 16S ribosomal DNA sequencing. Applied and Environmental Microbiology 67(4): 1565–1574.

Zelles, L., Bai, Q. Y., Beck, T. and Beese, F. 1992. Signature fatty acids in phospholipids and lipopolysaccharides as indicators of microbial biomass and community structure in agricultural soils. Soil Biology and Biochemistry 24(4): 317–323.

Zelles, L., Rackwitz, R., Bai, Q. Y., Beck, T. and Beese, F. 1995. Discrimination of microbial diversity by fatty acid profiles of phospholipids and lipopolysaccharides in differently cultivated soils. pp. 115–122. *In*: Collins, H. P., Robertson, G. P. and Klug, M. J. [eds.]. The Significance and Regulation of Soil Biodiversity. Developments in Plant and Soil Sciences. Vol. 63. Springer, Dordrecht.

CHAPTER 2

Methods for Isolation and Identification of Microorganisms

Asmaa Missoum

Introduction

The German physician Robert Koch, who was the first to invent solid media for bacteria, quoted a statement "The pure culture is the foundation for all research on infectious disease", since it is the basis of cultivating, isolating, and identifying microorganisms (Rodrigues et al. 2016). Other pioneer scientists include Anthony Van Leeuwenhoek, the first who built powerful microscopes to observe bacteria and protozoa, Richard Petri, who designed the Petri dish for cultivation of microorganisms on solid media, and Hans Christian Gram, who developed a staining technique to distinguish between Gram positive and Gram negative bacteria (Aiyer 2018, Sandle 2016). By definition, a pure culture is a collection of cells that have emerged from a single cell. Because in nature microorganisms live in mixed populations, it is possible to isolate those that exist in small numbers compared to others using certain enrichment media, as well as providing suitable environmental conditions for its growth. This was initially developed by the Russian microbiologist Sergei Winogradsky, when he succeeded in isolating nitrogen fixing bacteria with the help of Martinus Beijerinck (Kurtböke 2017, Aiyer 2018). For example, *Mycobacterium tuberculosis* can be isolated inexpensively using blood agar, since it is a pathogenic bacterium taken from blood samples of tuberculosis patients. Thus, this selective media, which imitate the initial environment where *M. tuberculosis* acts as a host, facilitate its cultivation for analysis and diagnosis (Drancourt and Raoult 2007, El Sahly et al. 2014). In addition to bacteria, there are other microorganisms that cause diseases such as fungi, protozoa, and viruses. Hence, the need to isolate and identify these microbes is a fundamental part of the disease diagnosis and prevention process, in which it plays a major role in public health practices. This is achieved by the use of culture, microscopy, and sometimes serology, as they are the most

Paris-sud université, UFR Sciences, 15 Rue Georges Clemenceau, 91400 Orsay, France.
Email: amissoum93@gmail.com

commonly used methods in all research laboratories (Rodrigues et al. 2016). The necessity of cultivating and characterizing microscopic agents has great benefits in biotechnology (exploiting cellular processes to make useful products), pharmacology (developing antibiotics from bacteria), and in environmental sciences (monitoring and preventing pollution). Despite these few mentioned examples, on the other hand, culture based systems have been challenged by the fairly recent progress of molecular based systems (Pankaj and Sharma 2018, Patra et al. 2018) that will be discussed more in this chapter.

Culture dependent identification

Streaking

After preparing the petri dish with an appropriate sterile medium, a sterile loop is used to transfer a drop from a bacterial suspension in a broth culture, and is then streaked from side to side across the agar's surface (Fig. 1) until covering about one third of the plate's diameter (Silva et al. 2012). This can be also carried out using a straight needle or a bent glass rod. After flaming the needle, it is streaked at the right angle to and across the previous streak. The procedure is repeated in such a way that the third streaking is parallel to the initial one. This helps to drag microorganism out in a long line from the first streak, thus thinning the colonies. These can be examined after 24 hours of incubation at 37°C (Keys et al. 2013).

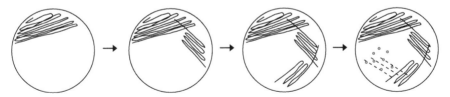

Fig. 1. Four quadrant streak plate method.

Plating methods

The plating method can be done in two different ways: pour plate method and spread plate method. In the pour plate method, colonies will grow inside and on the medium as the inoculum is mixed with melted agar, which is cooled to about 45°C, and left to solidify before incubation (Bielawska-Drózd et al. 2018). Pour plate method has few disadvantages such as loss of heat-sensitive organisms' viability and obligate aerobes in deep agar, time consuming preparation, and difficulty in examining embedded colonies (Meir-Gruber et al. 2016), whereas in spread plate method, 100 µL of the inoculum from broth media is spread across a solid agar in a Petriplate using a sterile spreader (Sanders 2012).

Serial dilution

Serial dilution is used when microorganisms cannot be simply isolated by plating and streaking methods. Serial dilutions from 10^{-1} to 10^{-10} were prepared (Yuan et al. 2018). After that, an inoculum is taken from each tube and plated on sterile solid media in Petriplates. This will allow isolated growth of microorganisms and give an opportunity to cultivate them in pure cultures (Yan et al. 2015).

Axenic media

In some cases, it is almost impossible to isolate certain species unless using a specific media type in addition to particular cultivation conditions. Therefore, enriched media are used to selectively grow microorganisms. The use of axenic media allowed a number of unrecognized gram-positive bacteria in the past, such as *Staphylococcus* species, to be isolated, generally from patients' blood or body fluids/pus (Houpikian and Raoult 2002). Other examples include the isolation of new enterohemorrhagic *Escherichia coli* serotypes by means of MacConkeys-sorbitol agar using stools from hemolytic-uremic syndrome patients (Houpikian and Raoult 2002), *Haemophilus influenzae* and *Neisseria gonorrhoeae* using chocolate agar from epiglottitis and phayingitis patients, respectively, in addition to *Corynebacterium diphtheriae* using Tellurite serum agar from diphtheria patients (Atlas and Snyder 2006). Moreover, recent findings show that Acidified citrate cysteine medium-Defined (ACCM-D) improved the yield and viability of *Coxiella burnetii* compared to previously used axenic culture media. Therefore, the assessment of this pathogen's nutritional requirements and its intracellular parasitism can be easily achieved using ACCM-D, as it enables to accurately define medium constituents for further metabolic studies (Sandoz et al. 2016).

Morphological characterization

Bacterial pure colonies differ vastly in terms of elevation, margin, and shape, illustrated in Fig. 2, in addition to surface characteristics (shiny or dull) and optical properties (translucent or opaque). Diameters also vary within species and some even produce pigments (Leboffe and Pierce 2011). Using light microscopy, cell morphology can

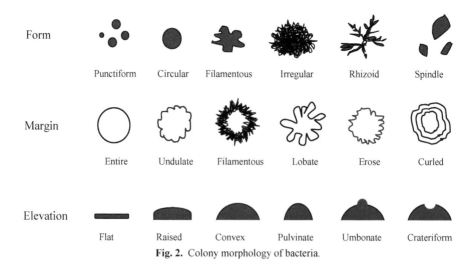

Form	Punctiform	Circular	Filamentous	Irregular	Rhizoid	Spindle
Margin	Entire	Undulate	Filamentous	Lobate	Erose	Curled
Elevation	Flat	Raised	Convex	Pulvinate	Umbonate	Crateriform

Fig. 2. Colony morphology of bacteria.

be also examined and it includes cell shape (rods, cocci, or spirillum), Gram stain reaction (positive or negative), organization (single, double, or chains), presence of endospores (terminal, sub-terminal, or central), and flagellation (peritrichous or polar). Furthermore, the use of scanning and transmission electron microscopy allows higher detail resolution and ultrastructural examination, respectively. However, regardless of the significant advantages, electron microscopy is expensive, time consuming, and requires highly experienced staff and sufficient knowledge of histology (Backert et al. 2018, Srinivasan et al. 2016).

Serological characteristics

Bacterial species can be classified on account of their characteristic antigens. These are molecules that trigger the immune system to produce antibodies and eventually interact with them. The specific interactions can be evaluated by several methods such as precipitation, agglutination, and ELISA. The most common found antigens on bacterial cells are somatic, flagelar, and capsular. Antigenically different strains within species are known as serotypes (Zourob et al. 2008). In a study conducted in Belgium recently, anti-*Borrelia* IgG antibodies were detected in 21.6% of the forest workers as a result of tick bites. The sero-prevalence of *Borrelia burgdorferi* was tested using ELISA method, which showed good specificity of 90% with a high sensitivity of 100% for the detection of *B. burgdorferi* (De Keukeleire et al. 2018).

Biochemical methods

The identification of bacteria is dependent on their ability to ferment sugars as well as on other functions of a variety of enzymes that are present in them. As mentioned earlier, some species require specialized media in order to be isolated first. This is to adapt certain nutritional requirements under investigation. Therefore, it is necessary to use the appropriate media when conducting these tests in order to insure positive reactions. Moreover, it is also important to check for impurities and the deterioration of the reagents (Harrigan and McCance 2014). Table 1 below summarizes a limited selection of experiments which are generally useful in the identification of bacteria.

It is also important to mention that researchers used *Bergey's Manual of Systematic Bacteriology* as a primary source of reference which categorizes and classifies bacterial species, based on recorded observations and revisions in taxonomy of microorganisms except fungi, protozoans and viruses (Pommerville 2018). Analytical profile index (API) is an example of a modified biochemical test that is alternatively used to facilitate the detection of pathogenic and fastidious bacteria belonging to *Enterobacteriaceae* and other families. API has different strips. For instance, API 20NE consists of a plastic strip that holds 20 small wells and each of these has dehydrated medium specific to one biochemical test. After incubation, the twenty test results will be interpreted using a given scale and the bacteria is identified to the species level (Sandle 2016). However, nowadays modern automated systems are used and developed in order to replace former time-consuming methods and traditional tests might result in false positive outcomes.

Table 1. Outline of the most commonly used biochemical tests in microbiology (Harrigan and McCance 2014).

Method	Principle	Result	Example of target microorganisms
Gelatin hydrolysis	Detect the ability of producing gelatinase (proteolytic enzyme)	Gelatin liquefied, even after exposure to cold temperatures	*Staphylococcus aureus, Bacillus anthracis, Bacillus cereus, Bacillus subtilis*
Casein hydrolysis	Detect the ability of producing extracellular enzymes, capable of degrading casein protein in milk as it is too large to enter bacterial cell membrane	Zone of hydrolysis around the line of growth in the skim milk agar	*Bacillus subtilis, Pseudomonas* sp.
Indole test	Detect the ability of splitting tryptophan to form indole compound via tryptophanase presence	Development of a blue color within 30 seconds	*Klebsiella oxytoc, Citrobacter koseri, Proteus vulgaris*
Hydrogen sulfide production test	Detect the ability of producing H$_2$S gas. Sulphur containing amino acids are decomposed	Formation of black precipitate on the medium	*Proteus vulgaris*
Urea hydrolysis test	Detect the ability of hydrolyzing urea into ammonia and carbon dioxide via urease presence	Slant color changes from light orange to magenta	*Helicobacter pylori, Proteus mirabilis*
Nitrate reduction test	Detect the ability of producing nitrate reductase that hydrolyzes nitrate (NO$_3^-$) to nitrite (NO$_2^-$)	Formation of a red precipitate after adding sulfanilic acid and α-naphthylamine. No color change after adding zinc powder	*Pseudomonas aeruginosa, Escherichia coli*
Starch hydrolysis test	Detect the ability of producing exoenzymes (α-amylase and oligo-1,6-glucosidase) that hydrolyzes starch	Zone of hydrolysis around the line of growth after adding iodine	*Bacillus megaterium, Bacillus subtilis*
Carbohydrate fermentation test	Detect the ability to ferment carbohydrates producing gas and/or acid	Acid production: broth turns yellow. Gas production: bubble in inverted Durham tube	*Escherichia coli, Clostridium pasteurianum*
Oxidase test	Detect the ability of producing cytochrome c oxidase (enzyme of the bacterial electron transport chain)	Development of indophenols showing dark purple-blue color within 10 seconds	*Legionella pneumophila, Moraxella* sp*., Neisseria* sp.
Methyl red test	Detect the ability of performing mixed acids fermentation from glucose	Media turns red after addition of methyl red	*Escherichia* sp*., Salmonella* sp*., Proteus* sp.

Voges-Proskauer test	Detect the ability of producing acetylmethyl carbinol from glucose fermentation	Development of red color after 15 minutes (diacetyl)	*Enterobacter aerogenes, Streptococcus mutans*
Citrate test	Detect the ability of using sodium citrate as the only source of carbon and ammonium salt as the only source of nitrogen	Development of blue color (alkalinization) with visible growth on slant surface	*Klebsiella pneumoniae, Salmonella* sp., *Edwardsiella* sp.
Tributyrin hydrolysis test	Detect the presence of lipase that hydrolyzes lipids	Zone of hydrolysis around the line of growth on tributyrin agar	*Pseudomonas aeruginosa*
Catalase test	Detect the presence of catalase that hydrolyzes hydrogen peroxide into water and oxygen	Production of copious bubbles	*Staphylococcus aureus, Corynebacterium diphtheriae, Burkholderia cepacia*
Coagulase test	Detect the presence of coagulase that is involved in clotting process	Formation of fibrin clot of any size	*Staphylococcus aureus, S. schleiferi, S. delphini*

Physical tests

Because microorganisms live in diverse habitats, each species surely has its own physiological characteristics that distinguishes it from others. This means that, for example, some bacteria live better or only in environments that are abundant in higher salt concentrations; these are known as halophiles (Gupta et al. 2014). Microorganisms that are adapted to such extreme ecological conditions are called extremophiles, and main reasons behind this are the remarkable enzymes along with various mechanisms in their cells (Horikoshi et al. 2011). Other types include psychrophiles (cold-loving) and thermophiles (heat loving), in addition to acidophiles that thrive in pH closer to 0, and alkaliphiles that live in conditions having pH higher than 10 (Rampelotto 2013).

In order to determine physiological properties of an isolated strain, its growth at different pH levels, NaCl concentrations, and temperatures can be evaluated using the best medium suited for its nutritional requirements (Zourob et al. 2008). There are also metallophiles surviving in high concentrations of heavy metals, and thus these can be identified by evaluating their growth in media incorporated with heavy metals. Barophiles, on the other hand, live in environments exposed to higher hydrostatic pressures such as deep oceans (Rampelotto 2013). They can be cultivated by diluting mud samples with Marine broth and inocuating on marine agar medium. Then, the plates are incubated at different atmospheric pressure values and growth is observed after a period of time (Horikoshi and Tsujii 1999). Although these methods remain the most commonly used to identify the physiological properties for all types of microorganisms, whether extremophiles or not, scientists continue to modify new methods in order to facilitate further research.

Susceptibility to antibiotics

This method is used for some bacteria that naturally resist anti-bacterial agents excreted from other competing microorganisms. These have particular cellular mechanisms responsible for this kind of defense, which are lacking in other species (Rahman et al. 2017). For that reason, it is a distinguishing characteristic that can be tested culturally using disc diffusion method. The discs, each containing a different type of an antibiotic, are placed on agar surfaces that are scabbed with certain bacterial broth. Then, the zones around the discs are observed after incubation and the susceptibility profile for that specific bacterium is evaluated (Fatima and Mussaed 2018, Tarale et al. 2015, Haghshenas et al. 2017). With the aid of molecular methods, this profile can be further fully presented as there are some microorganisms, especially pathogenic ones, which cannot be cultured easily in the laboratory such as *Prevotella* species (Rahman et al. 2017, Hahn et al. 2018).

Isolation and identification of anaerobes

There are types of bacteria that are unable to grow when there is 18% oxygen and 10% carbon dioxide in an atmosphere are known as anaerobes and lack enzymes, such as superoxide dismutase, that eliminate harmful oxygen species in their cells. The best way to take anaerobe-containing specimens is via use of sterile needles and syringes (Gajdács et al. 2017). Physically, McIntosh-Filde's jar is broadly used where inoculated plates are placed inside. Then, after sealing the chamber, the air inside is evacuated and replaced with other gases, followed by an incubation of 48 hours at 37°C. Because this requires costly apparatus and a particular vacuum pump, GasPak systems are more conveniently used nowadays in laboratories (Fang and Zhang 2015). In addition, the culturing anaerobic bacteria depends heavily on nutritionally rich media such as Holman-broth, thioglycolate-broth, and Schaedler blood agar. Once isolated, the anaerobes can be identified by all previously mentioned methods as well as other molecular methods (Gajdács et al. 2017).

Culture independent

Nucleic acid techniques

Many microorganisms cannot be cultured nor identified by culture-based tests in the laboratory. This includes more than 99% of soil bacterial species and several pathogens taken from human specimen. Out of the large number of reasons, why is that researchers are failing to replicate the necessary aspects of their environment. For this reason, nucleic acid-based methods are utilized instead to identify the DNA or RNA of these target microbes (Stewart 2012). The main principle of these techniques is based on nucleic acid hybridization, in which single stranded DNA or RNA molecules pair with complementary DNA or RNA and anneal to them under suitable physiological conditions (Aiyer 2018).

Sequencing methods

Bacterial genomes encode for all functions that are required for their survival. In terms of nucleotides' composition, they vary significantly between species. Within a DNA or a RNA molecule, the specific order of nucleotides can be determined using various technologies as they are very informative macromolecules. Sanger sequencing is one of the oldest methods that was replaced by modern automated sequencing platforms such dye terminator, high-throughput DNA, and Next Generation Sequencing. This is also conducted for other microorganisms (Donkor 2013).

16S ribosomal RNA sequencing

In the bacterial ribosome, the 30S subunit consists of 16S ribosomal RNA, while the 50S subunit consists of 5S and 23S ribosomal RNAs. Thanks to 16S rRNA gene's hyper variable regions, each species has recognizable signature sequence that is helpful for bacterial identification. Therefore, ribosomal RNA sequencing, whether 16S or 23S, is utilized to identify new species that could not be cultured in laboratories as well as reclassifying them in phylogenetic trees (Wanger et al. 2017).

Phylogenetic analysis

After obtaining the unknown sequences, the next step is to find the closest and most analogous sequence of 16S rRNA gene using BLAST search in one of the universal bases for genomic data: NCBI (http://www.ncbi.nlm.nih.gov/), EzTaxon (http://ezbiocloud.net), EMBL (http://ebi.ac.uk), and SILVA (https://www.arb-silva.de/). Sequence data from EMBL and NCBI are unfiltered while EzTaxon and SILVA utilizes best quality sequences resulting in a more reliable outcome that is easier to interpret. Subsequently, a phylogenetic tree is generated to give an insight into where the unknown is positioned (Kurtböke 2017).

Polymerase chain reaction (PCR)

Simply known as PCR, it is a technique that was invented by Kary Mullis in 1980s. It involves the amplification of target DNA through a series of controlled thermal cycles: denaturing, annealing, and extension presented in Fig. 3. In the reaction tube, DNA template, primers, dNTPs, and DNA polymerase are all added in a suitable buffer solution. Variations of this method include real time PCR, multiplexed PCR, and reverse transcriptase, as well as other recently modified protocols (Wanger et al. 2017).

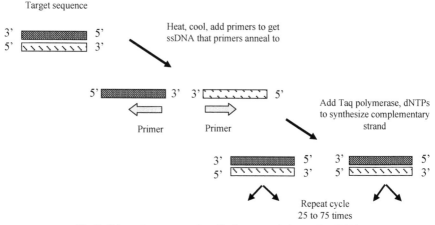

Fig. 3. Schematic representation of polymerase chain reaction (PCR).

Quantitative polymerase chain reaction (qPCR)

Quantitative polymerase chain reaction (qPCR), which is also known as real time PCR, quantifies the amount of gene product in the given sample during the cycling reaction. This is achieved as the machine sensors detect florescence from additionally produced DNA, and then plot it on a curve giving the corresponding quantity of the product after each cycle (Kralik and Ricchi 2017). It can also be used qualitatively to determine whether the DNA product is present or not by amplifying that specific DNA fragment, if present. This method facilitated the identification of *Leptospira* species in patients' urine samples in less than 2 hours compared to other reference tests with 100% accuracy. These species of bacteria cause Leptospirosis, an infectious disease that lacks pathognomonic signs and leads to organ failure (Esteves et al. 2018).

Multiplexed PCR

Multiplexed PCR uses several pairs of primers that will target numerous sequences in the target DNA producing different amplicons in a single reaction. This method was recently used to detect *Staphyloccocus aureus* (16s rRNA specific gene) and the mecA gene that encodes for oxacillin resistance directly from clinical blood samples (Rocchetti et al. 2018). It was found that 23.0% of the blood samples contained *S. aureus*, of them 50.6% possesses mecA gene. There was also 100% agreement between Multiplex PCR and the phenotypic identification in the study. Moreover, the method also detected the CoA gene distinguishing the coagulase negative *Staphyloccocus* species that were 77% of total isolates (Rocchetti et al. 2018).

Reverse transcriptase PCR

Reverse transcriptase PCR allows the amplification of mRNA transcripts from minute biological samples. Using the enzyme reverse transcriptase, complementary DNA transcripts are produced from the RNA templates and then multiplied by PCR (Sobhy et al. 2018).

Fluorescence quantitative real-time PCR (FQ-PCR)

Fluorescence quantitative real-time PCR, also simply abbreviated as FQ-PCR, has been beneficial in the detection of *Fusobacterium nucleatum* in colorectal cancer tissues in Chinese patients. This method has high probe sensitivity, high spectral technology precision, and high amplification efficiency. FQ-PCR is commonly used to detect pathogenic bacteria as it is able to distinguish between different types of infections, as well as whether they are subclinical or latent (Li et al. 2016).

Arbitrarily primed polymerase chain reaction (AP-PCR)

Arbitrarily primed polymerase chain reaction (AP-PCR), also known as random amplification of polymorphic DNA (RAPD), amplifies random DNA pieces of a target bacterium using single primers of arbitrary sequence. It does not require prior knowledge of the bacterium's DNA sequence. AP-PCR was applied to generate oral *Lactobacillus* species' DNA fingerprints for detection using single short primer and low stringency annealing temperature. Unfortunately, this method lacks reproducibility (Li et al. 2015).

Rep polymerase chain reaction (Rep-PCR)

Rep-PCR is another genomic DNA fingerprinting technique that has similar discriminatory power to RAPD. It is based on distribution of conserved repetitive DNA regions such as BOX and ERIC. Using BOXA1R primer, rep-PCR was proven to be the most reliable and effective method for typing and differentiating 21 strains of *Bifidobacterium* species. Compared to RAPD, rep-PCR necessitates higher stringency annealing conditions (Jarocki et al. 2016).

Typing methods

Ribotyping involves the isolation of total DNA and digesting it with specific restriction enzymes. Southern blotted have been performed prior to hybridization with 16S or 23S rRNA probes of target species. Since these genes have been highly conserved through evolution, certain pathogenic bacteria can be detected using radio-labeled probes that could be designed from known sequences (Quinn et al. 2011).

Restriction fragment length polymorphism (RFLP)

This DNA profiling method simply requires digesting the DNA sample and separating the restriction fragments on gel electrophoresis according to size, revealing specific blotting patterns. Because it focuses on polymorphism of one target gene, the remainder of bacterial genome is excluded (Papademas 2015).

Terminal restriction fragment length polymorphism (T-RFLP)

Terminal RFLP, also known as T-RFLP, is a variant of RFLP that is based on 16S rRNA gene variations. In this method, rRNA genes are amplified with fluorescently labeled primers, which enable the detection of 3' or 5' terminal fragments from all restriction fragments after restriction digestion. Following the separation of the fragments by gel electrophoresis, the fluorescently labeled terminal ones are then detected using an automated DNA sequencer (Papademas 2015).

Amplified ribosomal DNA restriction analysis (ARDRA)

Amplified rDNA restriction analysis is an extension of restriction fragment length polymorphism (RFLP) to the 16s gene. This approach involves the amplification of conserved regions at the ends of target 16s gene, followed by restriction enzymes digestion. Subsequently, the generated pattern represents the analyzed species. ARDRA was also considered as a reproducible and a reliable molecular tool to identify pathogenic bacteria (Jarocki et al. 2016).

Amplified fragment length polymorphism (AFLP)

This technique combines PCR with restriction digest analysis. When a large number of fragments is obtained after digesting the bacterial DNA, short adapter sequences are ligated to the free overhanging ends of the fragments (both must have complementary sequences). In order to amplify the target fragments, PCR primers are designed to bind to these adapter sequences but also have few additional bases at the 3' end that are template independent. This serves to reduce the number of PCR amplified products that are then separated by gel electrophoresis. Recently, fluorescent tags are attached to adapters in order to facilitate detection of amplicons through an automated DNA sequencer (Foley et al. 2011).

Multilocus sequence typing (MLST)

MLST is a recent subtyping technology that is based on DNA sequencing, which was initially developed to detect *Neisseria meningitidis*, a pathogenic bacterium that causes meningitis. This approach usually compares seven housekeeping genes, present in all strains of a particular species. After amplifying internal segments from these genes, they are sequenced automatically to detect polymorphisms (Jandova et al. 2016). Thus, the allelic variations at each of the house-keeping loci are assigned a sequence type (ST) number. In other words, each isolate or strain is distinguished by seven integers that can be compared with those of other strains of same species. The isolates' profiles of many microorganisms can be easily found at MLST global databases such as http://www.mlst.net/databases/default.asp and http://www.pubmlst.org/databases/. This is one of the advantages compared to conventional typing methods that involve comparing the sizes of DNA fragment on gels (Papademas 2015).

Loop-mediated isothermal amplification (LAMP)

LAMP is a widely used technique based on efficient amplification of nucleic acid sequences under specified isothermal conditions. According to its protocol, primers for specific bacterial detection and quantification can be designed to facilitate detection without the need for DNA extraction and purification. The detection takes less than 1 hour to complete and does not require expensive equipments or extensive staff training. In order to enhance sensitivity, concentrating cells is necessary before LAMP reaction (Soares-Santos et al. 2018). Recent findings showed that LAMP was beneficial in various clinical applications compared to other common laboratory methods. Using malB gene designed primers, it has detected *E. coli* in urine samples of urinary tract infection patients rapidly and efficiently than direct microscopic tests (Ramezani et al. 2018). Favorably, LAMP was also employed to detect myobacteria associated with pulmonary infections. Using a developed assay targeting rpoB gene and incorporated degenerate primers, it successfully identified twenty mycobacterial species from clinical respiratory tract samples, even more efficiently than qRT-PCR. However, occasional non-specific amplifications must be taken into account as a drawback (Grandjean Lapierre and Drancourt 2018).

DNA microarrays

Microarrays are remarkable technologies in which nucleic acid sequences are robotically deposited in arrays on micro-slides and fixed by forming chemical bonds. These nucleic acids hybridize with the mRNA probes on the array slides and because they are fluorescently labeled, they can be easily detected (Fig. 4). Consequently, thousands of sample DNA molecules bonded on the small chip or micro-slide are quantified and sequenced simultaneously (Walker and Rapley 2009). Similarly, genes associated with infection can be identified by hybridizing the isolated mRNA

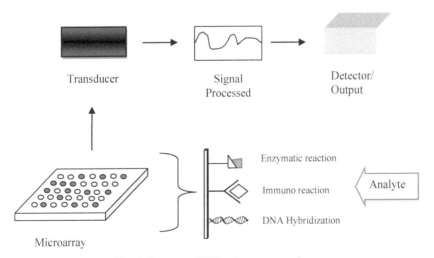

Fig. 4. Summary of DNA microarray procedure.

to a DNA microarray, which contains target gene sequences of a certain bacterial pathogen. These can be chemically synthesized amplicons obtained through PCR as mentioned in former step. This technology was also very useful in epidemic outbreaks as it has managed to carry out field tests of 300 samples per day detecting both natural and bioterrorist agents (Mirski et al. 2016).

Recently, the developed Phenotype MicroArray system, Biolog, was used to identify and characterize *Serratia* spp. This system consists of arranged well arrays (each tests a certain cellular phenotype) and an automated instrument, which monitors cells' response in the wells. Overall, this technology is composed of three components: EcoPlates microplates (physiological profiling evaluation), GEN III OmniLog ID System (isolate identification), and phenotypic microarrays, PM, technology (evaluation of antibiotics sensitivity). During plates' incubation, the tetrazolium dye is gradually reduced and recorded. In fact, the intensity of color change is directly proportional to the bacterial growth. The target strain was successfully identified as *Serratia marcescens ss marcescens*, and 16S rDNA sequencing gave the same results. Thus, Biolog is a reliable tool for detecting bacterial multi-antibiotic resistance (Chojniak et al. 2015).

Electrophoresis based methods

Pulsed field gel electrophoresis (PFGE)

PFGE is the type of gel electrophoresis where the direction of the applied electrical field can be changed periodically. When the chromosomal DNA sample is isolated and purified, it is digested with specific restriction enzymes. Then, the resulting fragments are resolved on the gel according to size. This method separates huge molecules of DNA effectively, which makes it useful in bacterial strain sub-typing. However, it requires costly equipments and cannot be automated as it is labor-intensive (Quinn et al. 2011).

Denaturing/temperature gradient gel electrophoresis (DGGE/TGGE)

DGGE uses PCR primers to amplify target DNA sequences, then separates the resulting amplicons in a polyacrylamide gel that contains a gradient of denaturing agent such as urea or formamide. This induces the DNA melting at various stages and the separated DNA sequences are of an identical length but different sequence (Xiong et al. 2018). TGGE is another variant of gradient gel electrophoresis whose principle is similar to DGGE, except that it relies on temperature to melt the DNA amplicons instead of chemical reagents (Benkeblia 2015). Although both techniques are used to analyze PCR product of the microbial 16S rDNA and 16S rRNA, the main drawback is that they require DNA and RNA extractions as well as sequencing steps after the gel separation. These laborious processes are burdened with errors and biases that might lead to the wrong identification of bacteria (Benkeblia 2015, Xiong et al. 2018).

Fluorescence in situ hybridization (FISH)

FISH is a cytogenetic method that uses fluorescence labeled probes to sense the absence or presence of particular sequences or regions on chromosomes. This is achieved by means of hybridization. Then, fluorescence microscopy is used to locate the fluorescent probe that is bound to target DNA. However, this method is expensive, time consuming, and could demand probes that are commercially unavailable (Bokulich and Bamforth 2018).

Immunological based methods

The enzyme-linked immunosorbent assay (ELISA)

ELISA is a rapid, inexpensive, and sensitive assay method that detects antigens or antibodies in a given sample. Because all microbes have unique antigens, they can be exploited as target molecules which can bind to capture antibodies in the assay. Sandwich ELISA identifies different epitopes at a time and was successfully used to detect *Bacillus cereus*, a common food poisoning pathogen (Zhu et al. 2016). Similarly, whole-EB ELISA quantified *Chlamydia trachomatis* specific antibodies IgG and IgA in genital secretions from female patients. This was achieved by the use of elementary bodies (EBs) from servers E and D. Although antibody optimization is difficult, this technique could help in designing an effective vaccine for *Chlamydia trachomatis* (Albritton et al. 2017). Recently, using Sao-M as a diagnostic antigen, modified protoplasmic antigen ELISA (PPA-ELISA) resulted in a more sensitive, specific, and rapid detection of *Streptococcus suis* for a large-scale epidemiological survey. However, it failed to distinguish vaccine strains of *S. suis* from the wild-type (Xia et al. 2017).

Field effect enzymatic detection (FEED)

FEED is an immuno-assay based-system that is employed to identify bacteria directly from blood samples without the need for culture enrichment. This operates by immobilizing a redox enzyme on the working electrode, using a sandwich immune complex. Next, an electric field is induced at the enzyme-electrode-solution interface by the gating voltage (V_G). The current between the redox enzyme and the electrode, known as the signal current, is then amplified by the V_G (Shi et al. 2018). One of the major drawbacks is that if a pathogen's antibodies are commercially unavailable, then new antibodies must be developed and tested.

Immunomagnetic separation (IMS)

IMS is a technology that uses antibody-and-antigen specific reaction as well as the magnetic beads' response for enrichment and separation. In other words, target bacterial cells are separated from others using magnetic bead. The surface of bead is coated with antibodies that can capture specific microorganism. IMS has a quick separation speed and is very specific and sensitive. For instance, this technology enabled a faster isolation of *F. tularensis* from blood cultures for antibiotic susceptibility determination. In addition, IMS was beneficial for *Y. pestis* as it exhibits rapid growth rates compared to *F. tularensis* (Aloni-Grinstein et al. 2017). Unfortunately, IMS was proven to have limitations such that it lacks pathogenic *E. coli* serogroup specificity. The difficulty in identifying these serogroups in pure cultures implies that they have low affinity for current commercial beads. Therefore, further refinements are required to improve IMS efficacy in the interest of clinical diagnosis and public health safety (Hallewell et al. 2017).

Colloidal gold immunochromatographic assay Colloidal GICA

For on-site, rapid detection of pathogenic *Spiroplasma eriocheiris* in crustaceans, a recent immunochromatographic test that uses gold nanoparticles was developed. GICA is based on sandwich immunoassay principle that involves specific combination linking polyclonal antibody and the pathogen on a nitrocellulose membrane (NCM). The visual appearance of both red lines at the test and control zones of the NCM confirms positive results within 15 minutes and without cross-reacting with other bacteria. In addition, gold nanoparticles with size of 40 nm and NCM with the pore size of 5 μm were chosen as the best factor for successful development of the test strip. The outcomes of the study were further confirmed by transmission electron microscopy (Liu et al. 2017). However, this strip test could be limited for only certain types of bacteria.

Spectroscopy based methods

The matrix-assisted laser desorption/ionization time-of-flight mass spectroscopy (MALDI-TOF)

MALDI-TOF is a mass spectrometry technique that identifies microbes to species level and has been widely used in diagnostic laboratories (Fig. 5). It is based on the precise determination of bacteria's protein mass that can be compared to stored profiles in software databases (Rodrigues et al. 2017). Using a fraction of a single pure colony, the extracted proteins are vaporized by ionizing laser to generate distinctive mass spectra. These have mass to charge ratio peaks that vary in intensities (Ge et al. 2016). The target proteins can be either conserved structural, ribosomal, or DNA

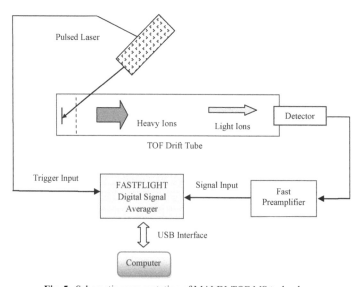

Fig. 5. Schematic representation of MALDI-TOF MS technology.

and RNA binding proteins. Their spectra are specific to genus, species, and even to subspecies. On the other hand, if the reference spectra of newly identified microbes are not on the database, then the characterization may not be feasible and the most similar spectra are used instead (Morka et al. 2018).

Nevertheless, this method is environmentally friendly as it does not require biohazard waste management unlike various biochemical tests that involve the use of microbial cultures. Moreover, MALDI-TOF improved antibiotic treatments by 18.9% for *Escherichia coli* and *Klebsiella pneumoniae*. This includes contributing to early patient's treatment as well as reducing the possibility of antimicrobial resistance (Ge et al. 2016). Comparison of MALDI-TOF key characteristics with other biochemical, spectrometric, and nucleic-based identification systems are summarized in Table 2 (Sakhno and Gunar 2016).

MALDI Biotyper CA System includes a huge library of clinically significant aerobic and anaerobic bacteria. Recently, it accurately identified 256 isolates from 17 *Corynebacterium* species. Furthermore, MALDI Biotyper CA System has optional processing techniques to enhance isolate characterization. Without affecting the reliability of the rest of the database, more isolates of a particular species can be added as part of library updates (Wilson et al. 2017).

Rapid evaporative ionisation mass spectrometry (REIMS)

REIMS is another mass spectroscopy method that, unlike MALDI-TOF, does not require time consuming cell extractions or preparative steps. This platform generates spectra of cell membranes' structural lipids, whose profiles can assist in detecting bacteria up to species level (Cameron et al. 2016). In REIMS analysis, microbial colonies grown on solid agar are subjected to an electric current by means of electro-surgical bipolar forceps. The particles that contain structural lipids as well as ionised metabolites will evaporate as a result of fast biomass heating. Consequently, the vacuum system will introduce these gas phase particles into the mass spectrometer (Bardin et al. 2018).

Raman spectroscopy

Raman spectroscopy is a label-free, non-invasive method that is capable of identifying single bacterial cells. It utilizes molecular vibrations to interpret chemical compositions and molecular structures of samples. The inelastic incident radiation scattering on the sample will interact with vibrating molecules, generating distinctive spectra that reflects molecular transitions. This technique is quantitative as there is a correlation between band intensity and species concentration. However, in the Raman spectrum, the signal response is quite low (Jehlička et al. 2014).

Surface-enhanced raman scattering (SERS)

Similar to conventional Raman spectroscopy, SERS is a spectroscopy method in which incident laser energy is scattered inelastically to generate spectral peaks. These are due to the molecule's vibrational forms. In this approach, molecules adsorbed on nanostructures or gold/silver roughened surfaces improve the Raman scattering. SERS is a widely used technology in many fields as it is rapid, sensitive and non-destructive. It also permits simultaneous multiplexed detection with high

Table 2. Comparison between different microbial identification methods (Sakhno and Gunar 2016).

Method	System	Database size	Time spent in detection	Manufacturer
Biochemical tests	Vitek 2 Compact	> 450 taxons	2–24 h	Biomerieux, France
	Biolog Microbial ID System	> 2500 species of bacteria, yeast and mold	4–26 h	Biolog Inc., USA
	API® and ID32	> 600 species of bacteria and yeast	18–48 h (2 h—for *Neisseria* spp., *Haemophillus* spp., *Moraxella catarrhalis*)	Biomerieux, France
Nucleic acid extraction, PCR amplification, and 16 rRNA based sequencing	MicroSeq™ Identification System	> 2300 bacteria species, 1100 fungi species	2–4 h	Applied biosystems, USA
Ribotyping	RiboPrinter® System	> 5700 patterns covering more than 180 genera, 1200 species	8 h	DuPont Nutrition & Health, USA
MALDI-TOF mass-spectrometry	Vitek MS	755 clinically important species	Less than 2 minutes	Biomerieux, France
	MALDI-Biotyper	> 2500 species (5600 strains) of microorganisms	Minutes	Bruker Daltonik GmbH, Germany
Fourier transform-infrared (FT-IR) spectroscopy	IFS-28B FT-IR spectrometer	730 bacteria strains covering 220 species out of 46 genera, 332 yeast strains covering 74 species out of 18 genera	Minutes	Bruker Daltonik GmbH, Germany
Fatty acid methyl ester analysis	MIDI Sherlock	> 2500 species including 700 environmental aerobic microorganisms, 620 anaerobic microorganisms and 200 yeasts species	Overnight	MIDI, USA

sensitivity. In case of identifying food borne pathogens, it is slowly replacing PCR and ELISA due to their time-consuming and complex pretreatments (Zhao et al. 2018). An example of a SERS substrate is, Ag SNP, a silver coated silicon nanopillar, which was used to detect mycolic acids that are mycobacteria species biomarker. The successful identification was validated by mass spectrometry and characterized three different forms of mycolic acids: αMA, keto-MA, and methoxy-MA in γ-irradiated whole bacteria. This is very helpful in differentiating between nonpathogenic and pathogenic *Mycobacterium* spp. (Perumal et al. 2018).

Fourier transform infrared spectroscopy (FTIR)

FTIR differs from Raman spectroscopy as it measures the distinct frequencies at which an isolate absorbs radiation as illustrated in Fig. 6. These correspond to the bonds' vibrational frequencies in the biomolecules of cells that are species-specific. Using biochemical fingerprints generated by FTIR spectra, a number of meat spoilage bacteria such as *Salmonella enteritidis* and *Listeria monocytogenes* were successfully identified. Although the signal response is higher than that in Raman's method, complex samples may result into misleading interpretations. Therefore, bacterial separation and purification as well as multiple trials are required (Grewal et al. 2015).

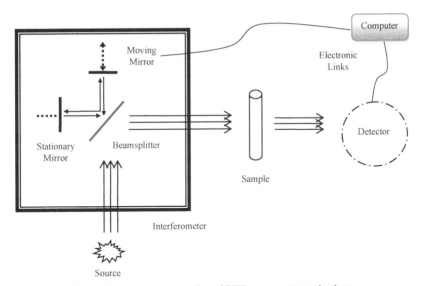

Fig. 6. Schematic representation of FTIR spectroscopy technology.

MIDI Sherlock

Phospholipid fatty acids (PLFA) and fatty acid methyl ester (FAME) analysis are common methods applied in microbial ecology to infer the structure of microbial communities in general. Since fatty acids are an important component in microbes' cell membranes, they can be used as biochemical marker to distinguish between main taxonomic groups. However, this does not permit species level detection (Benkeblia

2015). The two previously mentioned methods are also laborious as they require extraction and fractional methylation steps, in addition to gas chromatography analysis (GC). For this reason, newly developed systems such as MIDI Sherlock facilitate microbial identification by automatically designating and quantitating the GC peaks in an unknown sample. Then, the fatty acid composition profiles are compared to the stored MIDI database using Sherlock recognition software (Perez et al. 2017).

Conclusions

Microbiological isolation and identification techniques are enormously crucial in many scientific fields, whether in public health practices so as to detect pathogens in clinical diagnosis settings, or in biotechnological industries to develop new products exploited from microorganisms' biological processes, or even in green technologies to monitor and prevent environmental pollution. Although the conventional methods such as culturing and biochemical testing are still widely used at the present time, molecular based techniques such as PCR methods, microarrays, and MALDI-TOF are rapidly replacing them due to their preferred advantages that will prove very beneficial in near future along with new discoveries.

Acknowledgments

Author wishes to thank family and professors from Qatar University for their encouragements to write this chapter contributing to the book.

References

Aiyer, P. V. 2018. Introduction to Microbiology and Microbial Diversity (Vol. 1). New Delhi: Idea Publishing.

Albritton, H. L., Kozlowski, P. A., Lillis, R. A., McGowin, C. L., Siren, J. D., Taylor, S. N. et al. 2017. A novel whole-bacterial enzyme linked-immunosorbant assay to quantify *Chlamydia trachomatis* specific antibodies reveals distinct differences between systemic and genital compartments. PLoS ONE 12(8): e0183101. http://doi.org/10.1371/journal.pone.0183101.

Aloni-Grinstein, R., Schuster, O., Yitzhaki, S., Aftalion, M., Maoz, S., Steinberger-Levy, I. et al. 2017. Isolation of *Francisella tularensis* and *Yersinia pestis* from blood cultures by plasma purification and immunomagnetic separation accelerates antibiotic susceptibility determination. Frontiers in Microbiology 8: 312. doi:10.3389/fmicb.2017.00312.

Atlas, R. M. and Snyder, J. W. 2006. Handbook of Media for Clinical Microbiology. Boca Raton, FL: CRC Press/Taylor & Francis Group.

Backert, S., Tegtmeyer, N., Oyarzabal, O. A., Osman, D., Rohde, M., Grützmann, R. et al. 2018. Unusual manifestation of live *Staphylococcus saprophyticus, Corynebacterium urinapleomorphum,* and *Helicobacter pylori* in the gallbladder with cholecystitis. International Journal of Molecular Sciences 19(7): 1826. http://doi.org/10.3390/ijms19071826.

Bardin, E. E., Cameron, S. J. S., Perdones-Montero, A., Hardiman, K., Bolt, F., Alton, E. W. et al. 2018. Metabolic phenotyping and strain characterisation of *Pseudomonas aeruginosa* isolates from cystic fibrosis patients using rapid evaporative ionisation mass spectrometry. Scientific Reports 8: 10952. http://doi.org/10.1038/s41598-018-28665-7.

Benkeblia, N. 2015. Agroecology, Ecosystems, and Sustainability. Boca Raton, FL: CRC Press.

Bielawska-Drózd, A., Cieślik, P., Bohacz, J., Korniłłowicz-Kowalska, T., Żakowska, D., Bartoszcze, M. et al. 2018. Microbiological analysis of bioaerosols collected from hospital emergency departments and ambulances. Annals of Agricultural and Environmental Medicine 25(2): 274–279. doi:10.26444/aaem/80711.

Bokulich, N. A. and Bamforth, C.W. 2018. Brewing Microbiology: Current Research, Omics and Microbial Ecology. Norfolk, UK: Caister Academic Press. https://doi.org/10.21775/9781910190616.

Cameron, S. J. S., Bolt, F., Perdones-Montero, A., Rickards, T., Hardiman, K., Abdolrasouli, A. et al. 2016. Rapid evaporative ionisation mass spectrometry (REIMS) provides accurate direct from culture species identification within the genus *Candida*. Scientific Reports 6: 36788. http://doi.org/10.1038/srep36788.

Chojniak, J., Jałowiecki, Ł., Dorgeloh, E., Hegedusova, B., Ejhed, H., Magnér, J. et al. 2015. Application of the BIOLOG system for characterization of *Serratia marcescens* ss *marcescens* isolated from onsite wastewater technology (OSWT). Acta Biochimica Polonica 62(4): 799–805. doi:10.18388/abp.2015_1138.

Drancourt, M. and Raoult, D. 2007. Cost-effectiveness of blood agar for isolation of mycobacteria. PLoS Neglected Tropical Diseases 1(2): e83. https://doi.org/10.1371/journal.pntd.0000083.

De Keukeleire, M., Robert, A., Luyasu, V., Kabamba, B. and Vanwambeke, S. O. 2018. Seroprevalence of *Borrelia burgdorferi* in Belgian forestry workers and associated risk factors. Parasites & Vectors 11: 277. http://doi.org/10.1186/s13071-018-2860-2.

Donkor, E. S. 2013. Sequencing of bacterial genomes: Principles and insights into pathogenesis and development of antibiotics. Genes 4(4): 556–572. http://doi.org/10.3390/genes4040556.

El Sahly, H. M., Teeter, L. D., Musser, J. M. and Graviss, E. A. 2014. *Mycobacterium tuberculosis* bacteraemia: Experience from a non-endemic Urban Center. Clinical Microbiology and Infection: The Official Publication of the European Society of Clinical Microbiology and Infectious Diseases 20(3): 263–268. http://doi.org/10.1111/1469-0691.12298.

Esteves, L. M., Bulhões, S. M., Branco, C. C., Carreira, T., Vieira, M. L., Gomes-Solecki, M. et al. 2018. Diagnosis of human leptospirosis in a clinical setting: Real-time PCR high resolution melting analysis for detection of *Leptospira* at the onset of disease. Scientific Reports 8: 9213. http://doi.org/10.1038/s41598-018-27555-2.

Fang, H. H. and Zhang, T. 2015. Anaerobic Biotechnology: Environmental Protection and Resource Recovery. London: Imperial College Press.

Fatima, S. S. and Mussaed, E. A. 2018. Bacterial Identification and Drug Susceptibility Patterns in Pregnant and Non Pregnant UTI Patients. Singapore: Springer Singapore.

Foley, S. L., Chen, A. Y., Simjee, S. and Zervos, M. J. 2011. Molecular Techniques for the Study of Hospital Acquired Infection. Hoboken, NJ: Wiley-Blackwell.

Gajdács, M., Spengler, G. and Urbán, E. 2017. Identification and antimicrobial susceptibility testing of anaerobic bacteria: Rubik's cube of clinical microbiology? Antibiotics 6(4): 25. http://doi.org/10.3390/antibiotics6040025.

Ge, M., Kuo, A., Liu, K., Wen, Y., Chia, J., Chang, P. et al. 2016. Routine identification of microorganisms by matrix-assisted laser desorption ionization time-of-flight mass spectrometry: Success rate, economic analysis, and clinical outcome. Journal of Microbiology, Immunology and Infection 50(5): 662–668. doi:10.1016/j.jmii.2016.06.002.

Grandjean Lapierre, S. and Drancourt, M. 2018. *rpoB* targeted loop-mediated isothermal amplification (LAMP) assay for consensus detection of mycobacteria associated with pulmonary infections. Frontiers in Medicine 5: 332. doi:10.3389/fmed.2018.00332.

Grewal, M. K., Jaiswal, P. and Jha, S. N. 2015. Detection of poultry meat specific bacteria using FTIR spectroscopy and chemometrics. Journal of Food Science and Technology 52(6): 3859–3869. http://doi.org/10.1007/s13197-014-1457-9.

Gupta, G., Srivastava, S., Khare, S. and Prakash, V. 2014. Extremophiles: An overview of microorganism from extreme environment. International Journal of Agriculture, Environment and Biotechnology 7(2): 371. doi:10.5958/2230-732x.2014.00258.7.

Haghshenas, B., Nami, Y., Almasi, A., Abdullah, N., Radiah, D., Rosli, R. et al. 2017. Isolation and characterization of probiotics from dairies. Iranian Journal of Microbiology 9(4): 234–243.

Hahn, A., Burrell, A., Fanous, H., Chaney, H., Sami, I., Perez, G. F. et al. 2018. Antibiotic multidrug resistance in the cystic fibrosis airway microbiome is associated with decreased diversity. Heliyon 4(9): e00795. http://doi.org/10.1016/j.heliyon.2018.e00795.

Hallewell, J., Alexander, T., Reuter, T. and Stanford, K. 2017. Limitations of immunomagnetic separation for detection of the top seven serogroups of shiga toxin-producing *Escherichia coli*. Journal of Food Protection 80(4): 598–603. doi:10.4315/0362-028x.jfp-16-427.

Harrigan, W. F. and McCance, M. E. 2014. Laboratory Methods in Microbiology. Burlington: Elsevier Science.

Horikoshi, K. and Tsujii, K. 1999. Extremophiles in Deep-Sea Environments. Tokyo: Springer. doi:10.1007/978-4-431-67925-7.

Horikoshi, K., Antranikian, G., Bull, A. T. and Robb, F. T. 2011. Extremophiles Handbook (Ed., Vol. 1). Tokyo: Springer.

Houpikian, P. and Raoult, D. 2002. Traditional and molecular techniques for the study of emerging bacterial diseases: One laboratory's perspective. Emerging Infectious Diseases 8(2): 122–131. http://doi.org/10.3201/eid0802.010141.

Jandova, Z., Musilek, M., Vackova, Z., Kozakova, J. and Krizova, P. 2016. Serogroup and clonal characterization of czech invasive *Neisseria meningitidis* strains isolated from 1971 to 2015. PLoS ONE 11(12): e0167762. http://doi.org/10.1371/journal.pone.0167762.

Jarocki, P., Podleśny, M., Komoń-Janczara, E., Kucharska, J., Glibowska, A. and Targoński, Z. 2016. Comparison of various molecular methods for rapid differentiation of intestinal bifidobacteria at the species, subspecies and strain level. BMC Microbiology 16(1): 159. doi:10.1186/s12866-016-0779-3.

Jehlička, J., Edwards, H. G. M. and Oren, A. 2014. Raman spectroscopy of microbial pigments. Applied and Environmental Microbiology 80(11): 3286–3295. http://doi.org/10.1128/AEM.00699-14.

Keys, A. L., Dailey, R. C., Hitchins, A. D. and Smiley, R. D. 2013. Postenrichment population differentials using buffered *Listeria* enrichment broth: Implications of the presence of *Listeria innocua* on *Listeria monocytogenes* in food test samples. Journal of Food Protection 76(11): 1854–1862. http://doi.org/10.4315/0362-028X.JFP-13-089.

Kralik, P. and Ricchi, M. 2017. A basic guide to real time PCR in microbial diagnostics: Definitions, parameters, and everything. Frontiers in Microbiology 8: 108. http://doi.org/10.3389/fmicb.2017.00108.

Kurtböke, I. 2017. Microbial Resources: From Functional Existence in Nature to Applications (1st ed.). San Diego, CA: Academic Press.

Leboffe, M. J. and Pierce, B. E. 2011. A Photographic Atlas for the Microbiology Laboratory (4th ed.). Englewood, Co: Morton Pub.

Li, Y., Argimón, S., Schön, C. N., Saraithong, P. and Caufield, P. W. 2015. Characterizing diversity of *Lactobacilli* associated with severe early childhood caries: A study protocol. Advances in Microbiology 5(1): 9–20.

Li, Y. Y., Ge, Q. X., Cao, J., Zhou, Y. J., Du, Y. L., Shen, B. et al. 2016. Association of Fusobacterium nucleatum infection with colorectal cancer in Chinese patients. World Journal of Gastroenterology 22(11): 3227–33.

Liu, H., Xiu, Y., Xu, Y., Tang, M., Li, S., Gu, W. et al. 2017. Development of a colloidal gold immunochromatographic assay (GICA) for the rapid detection of Spiroplasma eriocheiris in commercially exploited crustaceans from China. Journal of Fish Diseases 40(12): 1839–1847. doi:10.1111/jfd.12657.

Meir-Gruber, L., Manor, Y., Gefen-Halevi, S., Hindiyeh, M.Y., Mileguir, F., Azar, R. et al. 2016. Population screening using sewage reveals pan-resistant bacteria in hospital and community samples. PLoS ONE 11(10): e0164873. http://doi.org/10.1371/journal.pone.0164873.

Mirski, T., Bartoszcze, M., Bielawska-Drózd, A., Gryko, R., Kocik, J., Niemcewicz, M. et al. 2016. Microarrays—new possibilities for detecting biological factors hazardous for humans and animals, and for use in environmental protection. Annals of Agricultural and Environmental Medicine 23(1): 30–36. https://doi.org/10.5604/12321966.1196849.

Morka, K., Bystroń, J., Bania, J., Korzeniowska-Kowal, A., Korzekwa, K., Guz-Regner, K. et al. 2018. Identification of *Yersinia enterocolitica* isolates from humans, pigs and wild boars by MALDI TOF MS. BMC Microbiology 18: 86. http://doi.org/10.1186/s12866-018-1228-2.

Pankaj and Sharma, A. 2018. Microbial Biotechnology in Environmental Monitoring and Cleanup. Hershey, PA: Engineering Science Reference.

Papademas, P. 2015. Dairy Microbiology: A Practical Approach. Boca Raton, FL: CRC Press.

Patra, J. K., Das, G. and Shin, H. 2018. Microbial Biotechnology: Volume 2. Application in Food and Pharmacology (Vol. 2). Singapore: Springer.

Perez, K. J., Viana, J., dos, S., Lopes, F. C., Pereira, J. Q., dos Santos, D. M. et al. 2017. *Bacillus* spp. isolated from puba as a source of biosurfactants and antimicrobial lipopeptides. Frontiers in Microbiology 8: 61. http://doi.org/10.3389/fmicb.2017.00061.

Perumal, J., Dinish, U. S., Bendt, A. K., Kazakeviciute, A., Fu, C. Y., Ong, I. et al. 2018. Identification of mycolic acid forms using surface-enhanced Raman scattering as a fast detection method for tuberculosis. International Journal of Nanomedicine 13: 6029–6038. doi:10.2147/IJN.S171400.

Pommerville, J. C. 2018. Fundamentals of Microbiology (11th ed.). Burlington, MA: Jones & Bartlett Learning.

Quinn, P. J., Markey, B. K., Leonard, F. C., Fitzpatrick, E. S., Fanning, S. and Hartigan, P. J. 2011. Veterinary Microbiology and Microbial Disease (2nd ed.). West Sussex, UK: Wiley Blackwell.

Rahman, M., Rahman, M. M., Deb, S. C., Alam, M. S., Alam, M. J. and Islam, M. T. 2017. Molecular identification of multiple antibiotic resistant fish pathogenic *Enterococcus faecalis* and their control by medicinal herbs. Scientific Reports 7: 3747. http://doi.org/10.1038/s41598-017-03673-1.

Rampelotto, P. H. 2013. Extremophiles and extreme environments. Life 3: 482–485. doi:10.3390/life3030482.

Ramezani, R., Kardoost Parizi, Z., Ghorbanmehr, N. and Mirshafiee, H. 2018. Rapid and simple detection of *Escherichia coli* by loop-mediated isothermal amplification assay in urine specimens. Avicenna Journal of Medical Biotechnology 10(4): 269–272.

Rocchetti, T. T., Martins, K. B., Martins, P. Y., Oliveira, R. A., Mondelli, A. L., Fortaleza, C. M. et al. 2018. Detection of the mec A gene and identification of Staphylococcus directly from blood culture bottles by multiplex polymerase chain reaction. The Brazilian Journal of Infectious Diseases 22(2): 99–105. doi:10.1016/j.bjid.2018.02.006.

Rodrigues, L., Abubakar, I., Stagg, H. and Cohen, T. 2016. Infectious Disease Epidemiology (1st ed.). Oxford: Oxford University Press.

Rodrigues, N. M. B., Bronzato, G. F., Santiago, G. S., Botelho, L. A., Moreira, B. M., Da Silva Coelho, I. et al. 2017. The matrix-assisted laser desorption ionization–time of flight mass spectrometry (MALDI-TOF MS) identification *versus* biochemical tests: a study with enterobacteria from a dairy cattle environment. Brazilian Journal of Microbiology 48(1): 132–138. http://doi.org/10.1016/j.bjm.2016.07.025.

Sanders, E. R. 2012. Aseptic laboratory techniques: Plating methods. Journal of Visualized Experiments: JoVE (63): 3064. Advance online publication. http://doi.org/10.3791/3064.

Sandle, T. 2016. Pharmaceutical Microbiology: Essentials for Quality Assurance and Quality Control. Amsterdam: Elsevier.

Sandoz, K. M., Beare, P. A., Cockrell, D. C. and Heinzen, R. A. 2016. Complementation of arginine auxotrophy for genetic transformation of Coxiella burnetii by use of a defined axenic medium. Applied and Environmental Microbiology 82(10): 3042–3051. http://doi.org/10.1128/AEM.00261-16.

Sakhno, N. G. and Gunar, O. V. 2016. Microbial identification methods in pharmaceutical analysis: Comparison and evaluation. Mathews Journal of Pharmaceutical Science 1(1): 001.

Shi, X., Kadiyala, U., VanEpps, J. S. and Yau, S.T. 2018. Culture-free bacterial detection and identification from blood with rapid, phenotypic, antibiotic susceptibility testing. Scientific Reports 8: 3416. http://doi.org/10.1038/s41598-018-21520-9.

Silva, N. D., Taniwaki, M. H., Junqueira, V. C., Silveira, N., Do Nascimento, M. D. and Gomes, R. A. 2012. Microbiological Examination Methods of Food and Water: A Laboratory Manual. London: CRC Press.

Srinivasan, L., Gurses, S. A., Hurley, B. E., Miller, J. L., Karakousis, P. C. and Briken, V. 2016. Identification of a transcription factor that regulates host cell exit and virulence of *Mycobacterium tuberculosis*. PLoS Pathogens 12(5): e1005652. http://doi.org/10.1371/journal.ppat.1005652.

Sobhy, N. M., Bayoumi, Y. H., Mor, S. K., El-Zahar, H. I. and Goyal, S. M. 2018. Outbreaks of foot and mouth disease in Egypt: Molecular epidemiology, evolution and cardiac biomarkers prognostic significance. International Journal of Veterinary Science and Medicine 6(1): 22–30. http://doi.org/10.1016/j.ijvsm.2018.02.001.

Soares-Santos, V., Pardo, I. and Ferrer, S. 2018. Direct and rapid detection and quantification of *Oenococcus oeni* cells in wine by cells-LAMP and cells-qLAMP. Frontiers in Microbiology 9: 1945. doi:10.3389/fmicb.2018.01945.

Stewart, E. J. 2012. Growing unculturable bacteria. Journal of Bacteriology 194(16): 4151–4160. http://doi.org/10.1128/JB.00345-12.

Tarale, P., Gawande, S. and Jambhulkar, V. 2015. Antibiotic susceptibility profile of *Bacilli* isolated from the skin of healthy humans. Brazilian Journal of Microbiology 46(4): 1111–1118. http://doi.org/10.1590/S1517-838246420131366.

Walker, J. M. and Rapley, R. 2009. Molecular Biology and Biotechnology (5th ed.). Cambridge, UK: Royal Society of Chemistry.

Wanger, A., Chavez, V., Huang, R., Wahed, A., Dasgupta, A. and Actor, J. K. 2017. Microbiology and Molecular Diagnosis in Pathology A Comprehensive Review for Board Preparation, Certification and Clinical Practice (1st ed.). Saint Louis, MO: Elsevier Science.

Wilson, D. A., Young, S., Timm, K., Novak-Weekley, S., Marlowe, E. M., Madisen, N. et al. 2017. Multicenter evaluation of the bruker MALDI biotyper CA system for the identification of clinically important bacteria and yeasts. American Journal of Clinical Pathology 147(6): 623–631. https://doi.org/10.1093/ajcp/aqw225.

Xia, X., Zhang, H., Cheng, L., Zhang, S., Wang, L., Li, S. et al. 2017. Development of PPA-ELISA for diagnosing *Streptococcus suis* infection using recombinant Sao-M protein as diagnostic antigen. Kafkas Universitesi Veteriner Fakultesi Dergisi 23(6): 989–996. http://doi.org/10.9775/kvfd.2017.18110.

Xiong, X., Hou, A., Yi, S., Guo, Y., Zhao, Z., Wu, Z. et al. 2018. Analysis of oral microorganism diversity in healthy individuals before and after chewing areca nuts using PCR-denatured gradient gel electrophoresis. Animal Nutrition 4(3): 294–299. http://doi.org/10.1016/j.aninu.2018.07.001.

Yuan, X., Bai, C., Cui, Q., Zhang, H., Yuan, J., Niu, K. et al. 2018. Rapid detection of *Mycoplasma pneumoniae* by loop-mediated isothermal amplification assay. Medicine 97(25): e10806. http://doi.org/10.1097/MD.0000000000010806.

Yan, Y., Kuramae, E. E., Klinkhamer, P. G. and Van Veen, J. A. 2015. Revisiting the dilution procedure used to manipulate microbial biodiversity in terrestrial systems. Applied and Environmental Microbiology 81(13): 4246–4252. http://doi.org/10.1128/AEM.00958-15.

Zhao, X., Li, M. and Xu, Z. 2018. Detection of foodborne pathogens by surface enhanced raman spectroscopy. Frontiers in Microbiology 9: 1236. doi:10.3389/fmicb.2018.01236.

Zhu, L., He, J., Cao, X., Huang, K., Luo, Y. and Xu, W. 2016. Development of a double-antibody sandwich ELISA for rapid detection of *Bacillus cereus* in food. Scientific Reports 6: 16092. http://doi.org/10.1038/srep16092.

Zourob, M., Elwary, S. and Turner, A. P. 2008. Principles of Bacterial Detection: Biosensors, Recognition Receptors, and Microsystems. New York: Springer.

CHAPTER 3

Application of Microorganisms

Govindan Rajivgandhi[1] *and Wen-Jun Li*[1,2,*]

Introduction

Microorganisms are very tiny sized organisms that live everywhere and cannot be seen through naked eyes (Koehler et al. 2019). They are present in very large numbers in the entire biosphere (Kallscheuer et al. 2019). Hypothetically, life is evolved from microbes (Dapurkar et al. 2017). So, microbes are simply called the chest of the world (Pessoa et al. 2019). The human body is also made up of 100 times more microbes than human cells. The live form survival is most competitive due to various environmental factors like stress, pH, temperature, nutrient depletion and various minerals (Li et al. 2019). There are approximately ten million microbes in every drop of water including genetically diverse bacteria, fungi, archaea, algae and protozoa, as well as viruses (Yang et al. 2008). In microscope, the microorganisms exhibit tiny rounded structures. These tiny creatures are called as a microorganism.

It can either be single cellular or multicellular (Ronald et al. 2001). They live in environments ranging from desert to muddy lands. Some microbes can grow on animals and others can grow freely (Kashyap et al. 2017). Among these, virus is different as it can grow only inside the host cells such as plant and animal cells, which are called obligate intracellular parasite. Microbes are unique in forms, but their activities are more diverse, bacteria that help humans digest food to the viruses that help plants resist heat, bacteria, viruses and fungi (Egert et al. 2016). If they are handling properly, they are providing more benefits in human life, which are the key components in food, medicine, agriculture and other areas. Every year, researchers are detecting new genera and new species of microorganisms to be applied in medical, pharmaceutical, food industry, agricultural industry and various other environmental research. In our lives, microbes contribute major role and are beneficial to us in many ways (Stout et al. 2017). On the contrary, a few microorganisms are harmful as they can cause infections. Based on the beneficial effects, the important uses of microbes

[1] State Key Laboratory of Biocontrol and Guangdong Provincial Key Laboratory of Plant Resources, School of Life Sciences, Sun Yat-Sen University, Guangzhou, 510275, PR China.
[2] Southern Laboratory of Ocean Science and Engineering (Guangdong, Zhuhai), Zhuhai 519000, China.
* Corresponding author: liwenjun3@mail.sysu.edu.cn

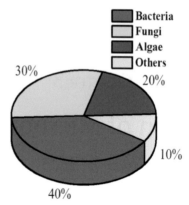

Fig. 1. Percentage of various microorganisms in all the fields.

in microbiology are classified into many categories including agricultural, food and dairy, pharmaceutical, medical, alcoholic beverages, enzyme technology, steroids and others (Jung et al. 2019). Based on the contribution, the microorganisms and their percentages are given in Fig. 1.

In the present decades, the rapid development of food, cosmetics, pharmaceutical, ecology and environment field by use of microbes is inevitable and cannot be ignored (Marienhagen et al. 2013). Microbes are very important to all the fields and the researchers are searching for alternative processes to find their major byproduct in a various sustainable way. From an industrial point of view, microorganisms are major candidates for the synthesis of biologically important products. Additionally, the microbes mediated byproducts are labeled as "natural", and also have low cost and low toxicity. For this fact, the industries are concentrating on microbes and success of its application is highly dependent on the availability of suitable microorganisms. It is related to increasing the demand in consumer market for natural products to substitute synthetic additives (Vitorino et al. 2017).

Application of microbes in various fields

The microbes contribute more to humans and environment. In addition, they extend to other fields also such as soil, agriculture, food and pharmaceutical industries (Fig. 2). Among the microbes, bacteria and fungi are the most predominant microbes that are more frequently used than other microbes in all the mentioned fields (Arezou et al. 2018). In soil and agricultural field, main role of the biosphere is as catalysts of biogeochemical cycle including carbon and nitrogen cycles as well as decomposition of dead plants and animals by various recycling processes (Shind et al. 2017).

In food industry, the microorganisms play an important role in butter, cheese, pasteurization, tyndallization and fermentation of various alcoholic processes. Likewise, microbes are help to maintain the water quality through physical and chemical properties such as temperature, pH, oxygen and salinity, which favor healthy fish and promote optimal growth. Similarly, one of the most important use of microorganism is in medical application. In medical field microorganisms act as an anti-toxin and vaccine for all the infections. Based on the highest usage of beneficial

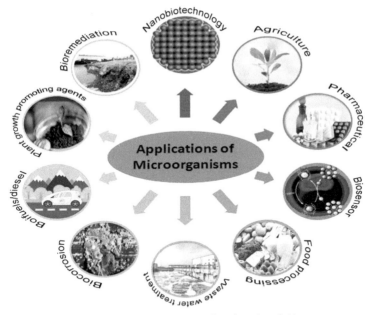

Fig. 2. Application of microorganisms in various fields.

effect, the bacteria and fungi have been chosen for this study. The importance of microbes in all the fields is given below in detail.

Application of bacteria and fungi in soil

Among the microbes, soil microbes are one of the important components for the sustainable production of healthy and organic foods (Motahhareh et al. 2018). In particular, the highest diversity of microbes containing rhizosphere is a chest of soil. In recent years, the agricultural biologist has contributed more to bacteria, fungi, algae and archaea because the soil bacteria plays an important role in regulating the soil fertility, nutrient cycle and plant diversity maintenance (Wentao et al. 2018). In the plant diversity, plant growth promoting bacteria is an important class in the soil and rhizobacteria that colonize root rhizosphere. Every time, agriculture is gained by naturally occurring plant growth promoting bacteria due to their beneficial effects on soil and crop productivity. In biotic or abiotic stresses, plant growth promoting bacteria help to enhance the plant growth. Among these, *Proteobacteria* and *Firmicutes* are the major bacteria that help soil (AndrAs et al. 2019, Wusirika et al. 2019). In addition, the phylum of *Firmicutes*, *Bacillus* sp. are the most predominant plant growth promoting bacteria including *Proteobacteria*, class Gamma proteobacteria such as the genus of *Pseudomonas*, *Acinetobacter*, *Serratia*, *Pantoea*, *Psychrobacter*, *Enterobacter* and *Rahnella*. In addition, the free living bacteria Burkholderia was frequently used for plant growth promotion (Constanza et al. 2019, Nur Sazwani et al. 2019, Olushola et al. 2019). In addition, the host mediated plant growth promoting comes under the family of *Fabaceae, Poaceae, Asteraceae, Brassicaceae, Asteraceae, Crassulaceae* and *Solanaceae*. In soyabean, pea and alfalfa growth, the legume family of *Fabaceae* contributed more for growth promotion. In this legume family, the symbiotic relationship between nitrogen-fixing endophytic bacteria and leguminous plants has been characterized very well (Rong et al. 2018). It is also connected with these plants related to free living *Pseudomonas* sp. and *Burkholderia* sp. as well as endophytic *Bacillus* sp. interacting with hyper accumulator plants (Liu et al. 2017).

For eradicating fungal contamination, the researchers have initiated many strategies for improving the crops (Zibo et al. 2018, Deniz et al. 2019). Among the strategies, biocontrol techniques are the most promising route to prevent the fungal contamination in pre and post harvested crops. In particular, some components of bacillus prevent the aflatoxins produced by fungi (Gonzalez Pereyra et al. 2018).

Bacillus is a Gram-positive rod; it has the ability to produce potential antifungal components and is tested mainly to control post-harvest fungal contamination (Mary-Anne et al. 2019). The *Bacillus* strains are well known reported bacteria which inhibit the fungal growth *in vitro* due to the antifungal antibiotics, especially the non-ribosomally synthesized cyclic lipopopeptide of the surfactin, iturin and fengycin groups (Deniz et al. 2019, Nur Sazwani et al. 2019, Olushola et al. 2019). One of the known potential biosurfactant is lipopeptides, an amphiphilic membrane active biosurfactant having the ability to inhibit fungal growth which can be used as biopesticides for plant and post-harvest protection (Misbahud et al. 2019). In addition, lipopolysaccharides secreting Bacillus spp., acts as an excellent antifungal agent which helps to increase the crop improvement in soil sample (Constanza et al. 2019).

For enhancing the crop yields, the inoculants of plant growth promoting rhizobacteria is an alternative solution via enhanced plant productivity, in a sustainable manner by direct and indirect mechanisms (Sanan et al. 2018). It delivers the nutrients to plants as well as produce some phytohormones such as auxin, gibberellins and cytokinins (Thiago et al. 2016), which forms part of direct plant growth promotion. In nitrogen fixation, *Azospirillum* sp. can produce and metabolize plant growth regulators, whereas *Pseudomonas* sp. with 1-aminocyclopropane-1-carboxylate (ACC) deaminase activity can regulate the tree imposed ethylene level in plants, inhibiting root elongation (Mercedes et al. 2017). *Bacillus* and *Pseudomonas* are involved in various biological controls in plant diseases.

Among the bacteria, *P. polymyxa* is an important Gram-negative bacterium which is involved in nitrogen fixation (Piotr et al. 2019). It is nonpathogenic and facultative anaerobic bacteria. It produces endospore. It is found in diverse ecological niches from a broad range of geographical locations. *P. polymyxa* is not only present in the rhizosphere of various crops, but also in marine sediments, forest trees, insect larvae and even clinical samples. During the last few years it has attracted global interest as this strain holds great promise for its ecological and biotechnological significance. So far, 150 species of *P. polymyxa* were reported from various soil samples (Bugra et al. 2019). The *P. polymyxa* is the highest cited strain in the research of plant growth promotion, nitrogen cycle, exopolysaccharides production and novel bioactive compound synthesis. It will help in biological control and for other industrial purposes.

Chitinase is an enzyme, which is synthesized by soil bacteria like bacillus, *Pseudomonas* and *Streptomyces* (Fenghua et al. 2019). It is used for biological control of agriculturally important plant pathogenic fungi, nematodes and insect pests. It has also become an emerging field of research used for pesticide management to sustain crop yields (Thiago et al. 2016). It is potentially used to plant pathogenic fungi, nematodes and due to synthesis of more toxicity metabolites synergism with Cry toxins in the control of phytophagous insect pests. The omnipresence of the substrate, its degrading enzymes and practical utility of both in agriculture and industrial sectors leads to the worldwide exploration studies on chitinolytic organisms (Nivien et al. 2018).

On the other hand, fungi are generally considered as oblicate aerobes traditionally (Mercedes et al. 2017). The fungi survives in ecology, water logged and poorly aerated soil including paddy soil. It is used widely in agricultural ecosystem. Fungi are more important microbes than bacteria due to their usage as soil fertility and crop improvement agents. In addition, it is used to compose the soil organic matter and maintain the ecosystem in acidic condition. It is a major source and regulator of energy flow and nutrient transformation in soil. The largest microbial diversity containing soil exhibited an excellent quality of soil, indicating excellent substrate use efficiency and higher nutrient availability (Piotr et al. 2019). Poor management and environmental difference may damage the microbial diversity and threatens soil health and productivity. Sometimes, it is more dynamic than physio-chemical characterization and is expressed as primary signal of soil degradation or amendment (Bugra et al. 2019).

Among the various fungi, *A. niger* is the most important fungus which produces phosphorus solubilizing components (Fenghua et al. 2019). It is used to help the increase of phytase, phosphatase and soluble phosphorus concentration in a very short period along with decreased concentration of phosphorus. It may be considered as utilization of fungi. Recent finding of Nur Sazwani et al. (2019) reported that the *Aspergillus* can increase the growth in high level after 50 h incubation under favorable condition. This statement agrees with previous report of Wusirika et al. (2019). For the solubilization and phytic acid for growth, a needed quantity of phosphorous become recovered to plant as well after the addition of fungal enzyme. Previous report of Constanza et al. (2019) also confirmed and reported that co-inoculation of fungal strains enhanced the root growth, shoot biomass and nitrogen in many crops. *Aspergillus* sp. is the best plant growth promoting fungi compared with bacteria, which travels long distance with phosphate solubilization. Interestingly, the phosphate solubilizing fungi and nitrogen fixing bacteria combine together in soil and reduce the drought stress condition in legume plants. Some important activities of bacteria and fungi in the soil industry are mentioned in the Table 1.

Application of bacteria and fungi in aquaculture

Aquaculture is an emerging and one of the fastest developing industries. It provides animal protein in high quality range. In 2011–2012, the protein production ranges

Table 1. The major role of bacteria and fungi in soil and agriculture industry.

S. NO.	Role	Microorganisms bacteria/fungi	Applications in various industries	References
1	Degradation	*Pseudomonas* sp., *Cupriavidus basilensis, Rhodococcus* sp.	In paper mills-lignin depolymerization and utilization. Also used for conversion of lignin to PHA. In waste water treatment, it is accumulated in lipid.	Motahhareh et al. 2018
2	Plant growth-promoting bacteria, biocontrol agents and plant stress homeoregulating bacteria	*Bacillus, Pseudomonas, Mycobacterium, Azospirillum, Agrobacteriu, Azotobacter*	They are used directly or indirectly to facilitate plant growth under biotic or abiotic stress conditions. The major important mechanisms of soil bacteria role are phosphate-solubilizing mechanisms and nitrogen fixation.	Constanza et al. 2019
3	Rhizosphere and endophytic bacteria	*Bacillus cereus* and Enterobactercloacae	Involved in crop production rate under metal stress condition and involved in heavy metal removal by metal bio-absorbents process.	Wentao et al. 2018
4	Biofertilizer	*Paenibacillus polymyxa*	This could reduce reliance on chemical fertilizers. In soil, it is used to improve the nitrogen fixation from atmosphere, increase phosphorus solubilization, iron acquisition and phytohormone production.	AndrAs et al. 2019
5	Chitinolytic bacteria	*Serratia* sp. and *Stenotrophomonas* sp.	Biocontrol agent referred to as pesticide through chitin degradation due to the production of chitinase.	Wusirika et al. 2019
6	Plant growth promoting bacteria	Gamma proteobacteria includes the genera *Pseudomonas, Acinetobacter Serratia, Pantoea,* and *Psychrobacter*	The symbiotic relationship between nitrogen-fixing and endophytic bacteria and sometimes production of secondary metabolites by root exudates.	Rong et al. 2018
7	Biodegradation	*Oxalobacteraceae* and *Ensifer* sp.	Degradation of di-(2-ethylhexyl) phthalate in plastic containing soil.	Zibo et al. 2018
8	Phosphate biofertilizer	*Aspergillus fumigatus* and *Aspergillus niger*	Improving the plant growth by degrading the phosphate solubilizing activity.	Deniz et al. 2019
9	Biofertilizers and biocontrol agents	*Trichoderma viride*	It played major role in ammonia volatilization from an agricultural system for plant growth improvement.	Mary-Anne et al. 2019
10	Phosphate-solubilizer	*A. niger*	Improving the plant growth through phosphate solubilization.	Fenghua et al. 2019

were 6.36–6.66 crore tones worldwide (Kannan et al. 2019). To meet this increasing demand in aquaculture, the farmers have faced challenge to increase growth rate and reduce risks of disease spread in aquatic animals. For this reason, the aquaculture practices have been altered and shifted from extensive culture to intensive culture (Anran et al. 2019). The food fish provides an average of one-fourth of the total animal protein intake of the world animal population. Worldwide, 70%, 65%, 60% 50% and 45% of the world aquaculture production was reported by China, Thailand, Germany, Russia and India (Thorsten et al. 2018). Among the countries, China is the most dominant, and the total production rose from 2.33 million tons in 1978 to 51.42 million tons in 2016, making China the only country in the world where aquaculture production exceeds the wild catch. However, the incidence of disease outbreaks in the farmed species has frequently evolved, leading to high mortalities and economic loss. For control of aquatic diseases, no vaccines are available. Vaccines will hardly be available soon due to the limitations in fish vaccine development. So far, some chemical additives and veterinary medicines and some antibiotics are only available for preventing the aquatic diseases (Schreier et al. 2019). For alternatives of physiochemical management, the microbes are used for better treatment by using recirculating aquaculture systems which requires a complete understanding of the biological components involved in water treatment (Paula et al. 2017). The inclusion of these systems contains biofilters of microbial communities comprised, whose structure, dynamics, and activities are responsible for system efficiency. In comparison, the chemical and physical factors are not a better choice to treat the aquaculture infections due to the changes in water as dissolved oxygen, temperature, salinity, and pH, which can be easily monitored and controlled. RAS biofilters are more complicated because their performance critically relies on the interactions of microbial communities in dynamic environments (Kuebutornyea et al. 2019).

For these reasons, probiotics have emerged as an alternative to antibiotics against various fish diseases. In aquatic environment, probiotics were frequently used successfully due to the shortfall of antibiotics (Per et al. 2017). Probiotic is defined as, "the active substances secreted by one organism that has an effective inhibition role against the growth of another organism". Based on the beneficial effect, some of the probiotics which are frequently used in aquaculture system are *Arthrobacter*, *Bacillus*, *Enterococcus*, *Lactobacillus*, *Lactococcus*, *Micrococcus*, *Pediococcus*, *Aeromonas*, *Burkholderia*, *Enterobacter*, *Vibrio*, *Pseudomonas*, *Rhodopseudomonas*, *Roseobacter* and *Shewanella* (Zhang et al. 2015, De et al. 2018). They are used to improve the growth and immunity of aquaculture species over the years. It also acts as a safe additive to improve the health of the host by enhancing growth, providing nutrients, modulating microbial colonization, improving immune responses, improving feed utilization, increasing digestive enzyme activities and digestibility, improving water quality and controlling diseases.

Recently, the Food and Agricultural Organization (FAO) and World Health Organization (WHO) announced probiotics as "live microorganisms which, when administered in adequate amounts, confer a health benefit on the host" (Goda et al. 2018). Some bacteria and fungi have majorly contributed in aquaculture field. They are listed in Table 2.

Table 2. The major roles of bacteria and fungi in the aquaculture industry.

S. NO.	Role	Microorganisms bacteria/fungi	Applications in various industries	References
1	Expression of respective genes	*Lactobacillus plantarum* subsp. *plantarum*	Inducing the interleukin-8 (IL-8) expression in the intestine and conferring protection against *Lactococcus garvieae* infection.	Schreier et al. 2019
2	Anti-bacterial role	*Lactococcus lactis* and *L. plantarum*	Prevents the multi drug resistant bacteria.	Thorsten et al. 2018
3	Probiotic and antagonistic activity	*Arthrobacter*	It is used to improve the growth rates, phenoloxidase activity, phagocytic activity and clearance efficiency of haemocytes of white shrimp in aquaculture water.	Zhang et al. 2015
4	Probiotics	*Bacillus cereus, Bacillus amyloliquefaciens, Bacillus baekryungensis, Bacillus licheniformis* and *Bacillus pumilus*	It is used in aquaculture field including diet improved growth performance, digestive enzyme activities, antioxidant function, immune responses, and disease resistance of fish.	Per et al. 2017
5	Probiotics	*Saccharomyces* sp. and *Saccharomyces cerevisiae*	It is used to improve the gut microvilli length and trypsin activity, decreased expression hsp70 in intestinal and anti-resistant against *A. hydrophila* infection.	De et al. 2018
6	Probiotic	*Hanseniaspora opuntiae*	It stimulates the immune system of sea cucumber and inhibits the *Shewanella marisflavi* growth by antagonistic activity.	Goda et al. 2018
7	Probiotics	*Rhodotorula* sp.	Used to inhibit the growth of the pathogen of *V. splendidus* in sea cucumber.	Kannan et al. 2019
8	Probiotics	*Lactobacillus* sp. and *Carnobacterium* sp.	Inhibition role against *Aeromonas* infections.	Anran et al. 2019

Application of bacteria and fungi in food industry

Food is very important for human beings which provides the nutrients for the survival of life (Qingyi et al. 2019). The role of microbes in food processing provides us with a variety of food products and sometimes, they too serve as a source of food. In food processing, various physical and chemical processes are applied to make food from raw materials. Among food processing, household and industrial food production are the major source of food preparation (Yuanmei et al. 2019). In food processing at household and industrial level, various microbes are involved in the food preparation and preservation. Homemade foods are prepared by the family members for their own consumption. Some of the bacteria and yeast are involved in the preparation of common foods like idly, and dosa from fermented rice.

In growing world, microbial foods have considerably increased due to increased demand and consumption of food for improved health and nutritional values (Liu et al. 2017). Apart from that, the microbes are not only used for good quality of food, they are also used for improving the bioavailability of nutrients, surface (texture), and to clear the dry skin. It helps in food safety by producing some inhibitory compounds, thus expanding the ability and safety of food product (Maeda et al. 2004). In fermentation, microbes are used to produce variety of substances, which have great biological role and biofunctionality to improve the food. Furthermore, it can be used in the treatment process of food contamination including probiotic, antimicrobial, antioxidant activities with polyglutamic acid, degradation of antinutritive compounds and some fibrinolytic activity. The microbial fermented food items promote health benefits to consumers (Liu et al. 2015).

In addition, all the fermented food products are also involved in the metabolic pathways in microbes and may produce enzymes, which can be used for conversion of chemical substitution to simple edible food with increased nutritive quality (London et al. 2015). The most important bacteria is lactic acid bacteria, involved in the fermentation of lactic acid yogurt and cheese. Notably, the bacteria play an important role in milk, fruit juice, plant product which helped to human health. In addition, it is used for the production of olives, pickles, sauerkraut, salami and sourdough bread in fermented vegetables and fermented meats, respectively. *Lactobacillus*, *Leuconostoc*, *Pediococcus* and *Streptococcus* are used in all foods and fermented food products. These bacteria are used to recover the lack of amino acid and vitamins (Yuanmei et al. 2019).

Another important type of microbe is yeast, which is used worldwide in the beverage and bakery industries for production of different fermented food items. It is available in two ways: either as a dry powder or as compressed cakes.

Yeast belongs to eukaryotic family of ascomycetes having high Vitamin B and proteins, generally called creamy yeast. The genus of the yeast is covered by *Saccharomyces*, *Candida*, *Torulopsis*, and *Hansenula* (Qingyi et al. 2019). In baking and brewer industries, the metabolic products are used frequently for food preparation. Currently, yeast has extended to various other divisions, including foods, drinks, pharmaceuticals, and modern compounds. For these process, *Saccharomyces cerevisiae*, *S. boulardii*, *S. cerevisiae* var. *boulardii*, and *S. carlsbergensis*, etc., are used (Qingyi et al. 2019). Similarly, molds are one of the important microbes for various food products including cheese, sausages, and soy sauce. In food industry, molds are inevitable and are also used in some innovative technology. In the food industry, the molds of *Aspergillus niger*, *A. nidulans*, *Rhizomucor miehei*, *Endothia parasitica*, and *Mucor* are in very high demand for production of various food products, food additives, enzymes, and organic acids (Yuanmei et al. 2019).

The beneficial role of bacteria in human life is designed as safe (Lynch et al. 2017). It synthesizes various metabolites and is used for life saving metabolic products such as organic acids, bacteriocins, mannitol, 1,3-propandiol, fatty acids, as well as exopolysaccharides. In food industry, microbial production of exopolysachharide, which is the most important metabolite, helps to prevent the surface contamination. Lactic acid bacteria are also involved in the production of EPS. They can be synthesized by external environment in the form of slime EPSs or adhere to the bacterial surface forming capsular EPS (Wei et al. 2019). In food industry, natural or safe thickeners, emulsifiers or stabilizers play a major role in increasing the structure of food. In addition, EPS is used for some additional beneficial role such as antioxidant, anticancer, immune modulatory, antibacterial and colonization of probiotic bacteria.

On the other hand, fungi are one of the prime factors in food industry to increase the food product yields, and are necessary when making products including bread, beer, cheese and coffee (Yuanmei et al. 2019). In the preparation of yogurt, chocolate and cheese, fungi have the probiotic property and also act as different flavoring agents. Also, fungi is a major ingredient in the enzyme manufacturing, organic acids, alcohols productions, etc. The unicellular yeast is used in the beer, wine and vinegar manufacture industries. The most important advantage of fungi in medical field is production of fruit niches, syrups, molasses, jams, jellies, wine, beer, etc. Worldwide, fungi like mushroom is a predominant delicious food and used in daily food. In the food industry, yeast, eukaryotic fungi and heterogeneous group of *Ascomycetes* and *Basidiomycetes* are the predominant microorganisms responsible for different range of fermented products such as fermented milk and dairy products including cheese, cereal and condiments (Liu et al. 2017).

Similarly, various types of lactic acid bacteria *Lactobacillus bulgaricus* and yeast *Saccharomyces cerevisiae* are involved in the preparation of curd from milk, yogurt and bread, respectively. Interestingly, microbes help to prepare the natural product of Toddy, a traditional drink without any alcoholic contamination. In alcoholic production, yeast is the most important product, which helps to improve the flavor, organic acid utilization and increases the nutritional properties. In addition, it is used to reduce the anti-nutritional factors and anti-toxins (London et al. 2015).

In industry, previously mentioned microbe *Saccharomyces cerevisiae* is used for production of ethanol with whiskey, brandy, beer, and rum. Protease is an enzyme, which has been synthesized by various microorganisms that are used in food preservation. In the tyndallization process, *Trichoderma* produced cyclosporins is a major immunosuppressive agent for preservation of food. Rotting grapes are used by molds for wine production. In addition, mushroom, spirulina and cyanobacterium are the most important fungi used as food source. Some fermented cereals and grains making yeast are used in beer production. Based on the beneficial role, some important bacteria and fungi and their role of food industry are presented in Table 3.

Table 3. The major roles of bacteria and fungi in the food industry.

S. NO.	Role	Microorganisms bacteria/fungi	Applications in various industries	References
1	Fermentation	*Lactic acid bacteria* and *Bifidobacteria* sp.	They are used in the food industry for heterofermentative or homofermentative metabolism.	Lynch et al. 2017
2	Fermentation	*Acetobacter aceti*	It is used for the production of vinegar.	Maeda et al. 2014
3	Growth improvement	*Acetobacter lovaniensis*	It is used in the growth of vegetable.	Lu et al. 2015
4	Ripening	*Acetobacter pasteurianus*	It is involved in chocolate manufacture industry.	London et al. 2015
5	Growth improvement	*Acetobacter tropicalis*	It is used for the production of coffee.	Yuanmei et al. 2019
6	Ripening	*Aspergillus oryzae*	It is used to prepare the soy sauce.	Qingyi et al. 2019
7	Tyndallization	*Ascomycetes* and *Basidiomycetes*	Used for the production of some food products.	Lynch et al. 2017
8	Fermentation	*Lactobacillus bulgaricus* and *Saccharomyces cerevisiae*	Preparation of curd from milk, yogurt, bread and prepare the natural product of Toddy, a traditional drink without any alcoholic contamination, respectively.	London et al. 2015

Application of bacteria and fungi in pharmacology

Pharmacology is a broad class of the field which describes the use of chemicals to treat and cure the diseases (Cintia et al. 2019). The microbial application in the pharmaceutical and health care environments is called pharmaceutical microbiology. Among the various industries, the most important microbial industry is pharmaceutical industry. It is used to develop different kinds of antibiotics for human, animal, veterinary, soil and medicinal use (Morena et al. 2016). A wide range of scope is always available in pharmaceutical microbiology. Also, the function of pharmaceutical microbiology is safe production of pharmaceutical, health care and medical devices. In the risk assessment process, it is used together with testing materials and monitoring environments and utilities.

In pharmaceutical industry, the microorganisms are used for development of antibiotics. Entire antibiotics are originally derived from microbial metabolism (Milagre et al. 2019). In the recent decades, for enhancing the drug efficiency, genetic manipulations with various methods have been enabled. In drug development, vaccines are very important antibiotic material that are synthesized by microorganisms. The production of various clinical related antibiotics is also discovered by various microorganisms. In addition, steroids are also derived from microorganisms (Hafiza et al. 2019).

It may be applied to the pharmaceuticals through the materials used for manufacture and various environmental sources during the process (Ramakrishnan et al. 2018). For the proliferation process, the microbes are needed in the pharmaceutical industry if conditions are favorable. All the infections of a patient not related to sterile product, whereas, a risk of the patients highly depends on non sterile products. The exact numbers and types of microbes are the key factors that need to be taken in to account.

To decrease the contamination, the most important key factor is minimizing the microbe's quantity and control the microbial growth. In pharmaceutical industry, the process runs with suitable control point, are in place and operating correctly, a frequent biocontamination control strategy needs to be in place. This can be performing on chest on controlling the virulence for contamination during production.

Pharmaceutical industry is a relatively small sector of manufacturing industry. In every hospital, a small or medium or big pharmacy unit is a major part of the hospital. It is an important sector among any other sector, highly innovative, regulated and more profitable. Currently, or in future, everyone can use the pharmaceutical products of pharmaceutical industry. The various ranges of the life enhancing drugs are used to treat most of the acute diseases, which have the ability to transform life saver molecules in chronically infected patient, debilitating diseases, also to proprietary medicines from the pharmacy to relieve the minor ailments. Globally, all the developed countries have allotted more money to pharmaceutical research than low income countries (François et al. 2019).

Among the various sources of pharmaceutical industry, microbes are a prime factor to develop a synthesis of wide range of pharmaceutical products, which helps in targeting practically any medical indication. It has enhanced the nature of anticancer drugs, vaccines, anti-infectious disease antibiotics and hormonal disorder vaccines, etc. (Madhurankhi et al. 2019). There is multi step complex route available for natural biosynthesis of endogenous molecules, some of which can be manipulated for the biosynthesis of foreign molecules. In this process, genetically modified microbes are used by recombinant technology or metabolically engineered by substantial alteration of their endogenous routes.

For the synthesis of chemical active pharmaceutical ingredients, fermentation process is the excellent route that relies solely on microbes with no equivalent in other biological systems including antibiotics, and secondary metabolites that are formed by bacteria and fungi acting as antimicrobial, anticancer and anti-infectious agents (Fadhile et al. 2017). These synthesized organic molecules are received by multi-step synthesis from their building blocks. Organic molecules are very complex in nature, potentially encompassing structure such as chiral centers, large stereo specific rings or unique conjugated double bond systems. Initially, the synthetic route is not only needed for development, but it is more time consuming and economic than fermentation process (Nadia et al. 2018). In new drug production, semi synthetic model fermentation method is more advantageous. Initially, the fermentation is used for the synthesis of natural molecules and then modified synthetically. It may reduce toxicity, increase potency and selectivity, and overcome bacterial resistance to traditional antibiotics.

For the naturally synthesized protein expression in microbial systems, fermentation is the sole source. Proteins are complex molecules and have high molecular weight. Their function and stability depend on the secondary and tertiary structure, as well as various post-translational modifications, mainly glycosylation (Wurong et al. 2017). One of the important characteristic of fermentation development is selection of strain and optimization, media and process development and finally scaling up to maximum productivity. In downstream process, different methods for extracting, concentration and purification of the product from a fermentation broth are utilized.

In the pharmaceutical industry, listed bacteria and fungi are frequently used. They are *Escherichia coli, Staphylococcus aureus* and *Streptomycetes, Streptomyces* sp., *Actinomyces* sp., *Nigrospora* sp., *Aspergillus* sp., *Saccharomyces cereviciae, Pichia pastoris* (Fadhile et al. 2017, Madhurankhi et al. 2019). Some pharmacology related bacteria and fungi are listed in Table 4.

Application of marine bacteria and fungi

The major portion of the earth is surrounded by marine environment and it is a chest of pharmacology (Yang et al. 2018). Hypothetically, life evolved from marine environment (Saurav et al. 2012). The life form survival is the most competitive in marine ecosystem (Magarvey et al. 2004). Marine microbes are the dominant life forms in oceans. There are approximately ten million microbes in every drop of surface seawater including genetically diverse bacteria, fungi, archaea, and protists, as well as viruses (Rashad et al. 2015). Among the microbes, bacteria are omnipresent, harmful, commensal and opportunistic. Life would be (nearly)

Table 4. The major role of bacteria and fungi in the pharmaceutical industry.

S. NO.	Name of the enzymes	Microorganisms bacteria/fungi	Applications in various industries	References
1	Biosurfactant	*Bacillus altitudinis, Bacillus amyloliquefaciens* and *Serratia marcescens*	They are used as emulsifiers and stabilizers in food industry, as formulations in cosmetic industry, as biocontrol agents in agriculture or in biodegradation and bioremediation in environmental protection system.	Cintia et al. 2019
2	Biosurfactant	*Rhizopus arrhizus* and *Fusarium fujikuroi.*	It is used as stabilizer in food industry.	Morena et al. 2016
3	Anti-oxidant activity	*Marinobacter* sp., *Bacillus* sp. and *L. delbrueckii*	They are used for the production of antioxidant compound and have scavenging DPPH free radicals which prove its potential to be applied in food, cosmetics and oil industries as safe, natural and biodegradable.	Milagre et al. 2019
4	Anti-oxidant activity	*Aspergillus oryzae* and *Glomus genus*	They are used for the degradation process of DPPH.	Wurong et al. 2017
5	Anti-microbial activity	*Bacillus* sp.	Anti-microbial activity against food pathogens.	Morena et al. 2016
6	Anti-microbial activity	*Cephalosporium* and *Fusarium solani*	They have the anti-microbial ability against Gram negative bacteria.	Fadhile et al. 2017
7	Anti-fungal	*Cladosporium delicatulum*	They inhibit the various fungal strains due to the production of fungal metabolites.	Nadia et al. 2018
8	Biological activity	*Streptomyces* sp. and *Nocardiopsis* sp.	They have excellent biological activity components including anti-microbial, anti-cancer and anti-viral activity.	Ramachandran et al. 2019, Rajivgandhi et al. 2018

impossible without bacteria in the oceans (Held et al. 2019). So far, the discovery of bacteria and their novel evolutionary metabolic descendant isolated from marine environment is estimated ≈ 10% (Magarvey et al. 2004). In this current scenario, the need for research, and focus on the novel source of organism followed by their bio-potential estimation are the primary concerns.

Among the marine microbe, actinomycetes are Gram positive filamentous bacteria with branched morphology, aerobic and unicellular with high percentage of G≡C (70%) in their genetic structure. ≈ 70% of the available antibiotics have been derived from these excellent secondary metabolites producer (Ramachndran et al. 2019). Actinomycetes associated with marine plant have been explored by many research groups worldwide (Matsumot et al. 2017). In recent decades, the taxonomic evolution of endophytic actinomycetes has emerged from marine rather than terrestrial nature (Fan et al. 2018, Dinesh et al. 2017). Among the microbial community, actinomycetes is the supreme secondary metabolites producer (Kemung et al. 2018). They are characteristic producers of growth promoting factors, hormones and potential antimicrobial components (Masand et al. 2015).

The excessive usage of antibiotics leads to development of multi drug resistant bacteria. On other hand, it leads to discovery of novel compounds of antibiotics from new sources such and marine microorganisms (Subramani et al. 2012, Ramachndran et al. 2018). Encouragingly, the rare group of actinobacteria, as non-*Streptomyces* or newl *Streptomyces* found in endophytic nature has already helped in the discovery of novel antimicrobials with unique chemical moieties, confirming that microbial natural products are still a promising source for drug discovery (Jose et al. 2016). The important role of marine actinomycetes against all the fields are exhibited in Fig. 3

To date, no comprehensive investigation has been performed on the cultural diversity of the actinobacteria associated with marine sources. Some attempt has been made to discover new taxa and their potential novel metabolites from endophytic actinobacteria with potential antibacterial and anticancer properties that play a role in controlled drug delivery. For example, non-*Streptomyces* genera such as *Micromonospora, Nocardiapsis, Actinomadura, Actinoplanes, Streptoverticilllium*

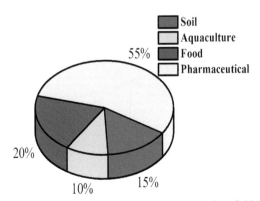

Fig. 3. Application of marine actinomycetes in various fields.

and *Saccharopolyspora* have been found to produce chemically unique antibiotics featuring potent biomedical activities such as abyssomicins, Pyrrolo [1,2-a] pyrazine-1, 4-dione, hexahydro-3-(2-methylpropyl), 1, 4-diaza-2, 5-dioxo-3-isobutyl bicyclo[4.3.0]nonane, rebeccamycin, macrotermycins, proximicins, and gerumycin, respectively (Li et al. 2017, Rajivgandhi et al. 2018).

Application of microorganisms in nanoparticle synthesis

Currently, nanotechnology is an emerging field that deals with nanosized materials (Xiaojia et al. 2019). The potential benefits of nanotechnology are widely recognized in all the fields including soil, agriculture, food, pharmaceutical, biology, physics and chemistry (Fig. 4). The building blocks of nanotechnology are nanoparticles, which are exhibited at various ranges between 1–100 nm (Tohren et al. 2019). Among the bulk materials, the tiny size and high surface area are totally different. In addition, nanoparticles and bulk materials also differ from each other due to the physical strength, reactivity, electrical conductivity, optical feature and magnetism. These nanoparticles are used in all the fields including energy, pharmaceutical, biomedical, cosmetics, textiles, food and agriculture. In the recent years, the nanoparticle research has concentrated on all the fields, particularly creating the revolution of agriculture, food and health sectors with the application of biosensors, plant growth promoters, food ingredients, drug delivery, pesticides and fertilizers (Mohamed et al. 2016). The synthesis of nanoparticles derives from chemical, physical and biological route. Compared to biological methods, physical and chemical methods are more toxic to all the fields including agriculture, food and health (Niladri et al. 2018). Therefore, synthesis of nanoparticle from biological route has exhibited excellent polydispersity, dimension as well as stability. The biological synthesis of nanoparticle using pH, temperature and pressure are cost effective and ecofriendly. Due to this reason, synthesis of nanoparticle from biological route especially microorganism is very important (He et al. 2019).

Worldwide, plant and microbes are exploited for synthesis of nanoparticles (Mohamed et al. 2016). Among the various microbes, bacteria, fungi and yeast are

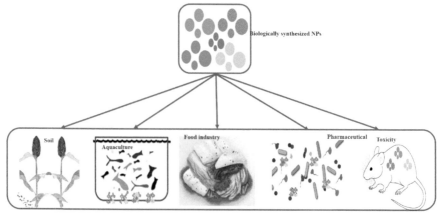

Fig. 4. Application of biosynthesized nanoparticles in various fields.

the most predominant microbes chosen for synthesis of nanoparticles due to the high growth rate, easy cultivation and their ability to grow in ambient atmosphere of temperature, pH and pressure. Based on the adaptability, microbial nanoparticles are synthesized by intracellular and extracellular route due to the enzymatic activities.

Among these, bacteria are an excellent nanoparticle producer due to the metal ion reduction. In the recent years, the synthesis of nanoparticles is concentrated on bacteria mediated interactive pathways responsible for metal ion reduction and their ability to precipitate metals on nanometer level (Tohren et al. 2019). Microorganisms are popular for nanoparticle synthesis due to ease of handling, minute size, diverse shape and less growth time. It exhibits low toxicity. On the other side, fungi are also one of the important microbes for nanoparticle synthesis. The fungi have various intracellular and extracellular enzymes in their nature, which are used for the synthesis of nanoparticles. It has been found that nanoparticles synthesized from fungi are mostly mono dispersed with well characterized size and shape, and also have different chemical compositions.

Here, important bacteria and fungi are *Actinobacteria, Bacillus* sp., *Pseudomonas* sp., *Klebsiella* sp., *Salmonella typhi* and *Fusarium oxysporum, Verticillium leuteoalbum, Collitotrichum* sp., *Alternaria alternate* and *Trichoderma viridea,* respectively, with some anti-bacteria and anti-cancer mechanisms (Niladri et al. 2018, Tohren et al. 2019). The other bacteria and fungi with their biological activities are also listed in Table 6.

Application of biosynthesized nanoparticles in soil

Nanoparticle is an excellent tool in agriculture in the form of nanofertilizer, nanopesticide, nanosensor which reported better result in sustainable agricultural practices (Nacide et al. 2019). Different nanoparticles are used in soil in laboratory and green house model (Table 5). Commercially available product of the nanoparticles is also available for using in the soil field (Rajendran et al. 2018). The commercially prepared nanoparticle for soil products are Nano-Gro™, Nano-AgAnswer®, Biozar Nanofertilizer, Nano Max NPK Fertilizer, Master Nano Chitosan Organic Fertilizer and TAG NANO; they are developed as plant growth regulator and immunity inducer, plant nutrition, NPK Fertilizer and Nano-Gro™ from respective countries of India, USA and Iran (Parada et al. 2019, Ralstonia solanacearum et al. 2019, James et al. 2017).

Application of biosynthesized nanoparticles in aquaculture

The excessive use of the antibiotics is a major reason to combat the bacterial infections. However, unregulated and unprescribed format of the antibiotics can also lead to emergence of bacterial resistance which is no longer effective on antibiotics (Arabinda et al. 2013). In mobile genome of both terrestrial and aquatic bacteria, elements have the resistant genes that can be exchanged between them with important impacts on livestock and human health. Importantly, tetracycline is the most frequently used antibiotic in fish forms for inhibiting the development of resistant genes in fish (Chandran et al. 2016). However, some bacteria are developed

Table 5. Importance of biosynthesized nanoparticles in soil.

S. NO.	Role	Microorganisms bacteria/fungi	Applications	References
1	Biofertilizers	*Bacillus*+Zinc	It is used to improve the soil fertility.	Tohren et al. 2019
2	Phytoremediation	*Streptomyces* sp.+Silver	It has phytotoxic effects and reduces the root and shoot biomass in *Sinorhizobium meliloti*.	Xiaojia et al. 2019
3	Phytoremediation	*Actinomycetes*+Gold	Increases the *Cinnamomum zeylanicum* for inhibiting the plant pathogens *Pseudomonas syringae*.	James et al. 2017
4	Simulator	*Aspergillus*+Copper	It increases the enzyme inhibition in soil bacteria	Parada et al. 2019
5	Crop improvement	*Crptococcus*+Cu-FAU nanozeolites	It is used to remove the ESKAPE pathogens.	Rajendran et al. 2018

to resist all kind of antibiotics including tetracycline. Particularly, *Aeromonas hydrophila* is the major organism that is resistant to broad spectrum antibiotics such as tetracycline, streptomycin and erythromycin. It produces skin ulceration, tail or fin rot and fatal hemorrhagic septicemia in fish. It commonly infects carp, gold-fish and silver catfish, leading to more economic losses in aquaculture (Eduardo et al. 2014). In addition, some of the microbes frequently create problems in fish forms including *Aeromonas salmonicida*, *Photobacterium damselae*, *Yersinia ruckeri*, *Vibrio*, *Listeria*, *Pseudomonas* and *Edwardsiella* (Mohamed et al. 2016).

For decreasing the antimicrobial resistant effect of fish pathogens, development of a potential alternative compound to replace the use of antimicrobials is the major concern. In this context, nanoparticle mediated drug delivery system in fish production form is an excellent alternative to fish pathogens. It has been developed to prolong and optimize drug administration and decrease the toxicity in fish form.

Application of biosynthesized nanoparticles in food industry

The food is our life partner to modern consumer, it led to increase in the demand for ready to eat or minimally processed foods. The consumer demand is for safe, fresh, healthy and shelf stable, convenient food using environmentally friendly methods (Niladri et al. 2018). Food is easily contaminated by microorganisms during the process and distribution including peeling, cutting and washing. If it is not handled properly, it could affect the public health. While handing and distribution of the food, continuous maintenance of desired temperature is sometimes difficult. The microbial contamination has led to the increase in various infections and poor nutrition associated with weaning foods. Sometimes, most of the food pathogens develop resistance against almost all the antibiotics that can lead to severe infections to healthy individuals. Prevention of resistant pathogens in food is the most important concern worldwide (He et al. 2019).

Nanotechnology in food is an emerging field and also new compared to the biomedical application. In food industry, it is frequently used during their cultivation, production, processing and food packaging (Thea et al. 2018). The use of solid based nanoparticles (NPs) has advantages over liquid based NPs due to the control release kinetics, nanoemulsions and coated products. These solid based NPS also have advantages over new techniques including nanosendors, tracking devices, targeted delivery, food safety and smart packaging (Khorasani et al. 2018).

Application of biosynthesized nanoparticles in pharmaceutical industry

The synthesis of nanoparticles from microbes exhibits several hypotheses. Particularly, silver, gold, zinc and copper are the most extensively studied nanoparticles in the pharmaceutical industry (Manjunath et al. 2019). In antimicrobial activity, silver nanoparticles release ionic silver via breakdown that inactivates the bacterial enzyme production by interacting with essential -SH groups. The released Ag ions is an active inhibitor for bacterial DNA replication, cytoplasmic membrane, lacking of adenosine triphosphate (ATP) and biofilm (Gharieb et al. 2018). Instead of silver, the biosynthesized gold nanoparticles are the most biocompatible and biodegradable. Nanoparticles are used in various biological applications, and can be used in photochemical and chemical modification by bioengineered method. It is also classified into different categories such as gold nano-cages, nanospheres, nano-needles, and nanorods. Likewise, zinc, copper, graphene and various other nanoparticles also have great potential against various biomedical applications (Table 6).

Table 6. Importance of biosynthesized nanoparticles in biomedical application.

S. NO.	Role	Microorganisms bacteria/ fungi	Applications	References
1	Antimicrobial activity	*Thermomonospora* sp.+Silver	Antimicrobial activity against urinary tract infections.	Hulkoti et al. 2014
2	Anti-microbial activity	*Penicillium chrysogenum*+Zinc	Antimicrobial activity against multi drug resistant bacteria.	Mohamed et al. 2016
3	Antioxidant	*Cladosporium cladosporioides*	It has more anti-oxidant activity.	Gharieb et al. 2018
4	Anti-cancer	*Cladosporium cladosporioides*	It has anti-candidal activity.	Manjunath et al. 2019
5	Anti-malarial	*Callistemon citrinus*-Gold	It has anti-malarial activity.	Thea et al. 2018

Conclusion

In human life, the microorganisms play a vital role, which help us to digest the food, decompose waste and are involved in various other life processes. Microorganisms have a more beneficial effect on humans than causing infections and damage to life. Microorganisms are classified in various categories based on the biotic processes; they are beneficial to soil, agriculture, food, pharmaceutical and various biotechnological industries. It is an inevitable source for sustaining life and our surrounding environment including soil, water and atmosphere. The application of microorganisms in soil industry can improve the crop of the plant and soil fertility during various metabolic cycles such as nitrogen and carbon cycles as well as decomposers in food chain of the soil. These organic cycles not only help human, but also play a significant role in plant growth promotions. Moreover, they are used to protect the environment and human health from the bio-remediation of harmful chemical and industrial wastes by degradation process. The application of microbes in the food industry has great benefits due to the fermentation and production of dairy products and this is the best example where microbes are beneficial to humans. In addition, application of microbes in the pharmaceutical and medicinal industry is the most important industries that are beneficial to human health. It also saves the economy because of more quantity of medicines being generated by various microbes. Microbes mediated products with complete understanding of the biological components are the best choice for all the fields including soil, agriculture, food, pharmaceutical and veterinary. Overall, the microorganisms are the heart of life on earth.

References

AndrAs, K., Zsuzsanna, N., Csaba, R., Balazs, V., Jozsef, K., Ildiko, P. et al. 2019. Monitoring of soil microbial inoculants and their impact on maize (*Zea mays* L.) rhizosphere using T-RFLP molecular fingerprint method. Applied Science and Ecology 138: 233–244.

Anran, W., Chao, R., Yanbo, W., Zhen, Z., Qianwen, D., Yalin, Y. et al. 2019. Use of probiotics in aquaculture of China—a review of the past decade. Fish and Shell fish Immunology 86: 734–755.

Arabinda, M., Snehasish, M., Ranadhir, B., UK, M., Surya, P. N. and Biplab S. 2013. Phytoextracts-synthesized silver nanoparticles inhibit bacterial fish pathogen *Aeromonas hydrophila*. Indian Journal of Microbiology 53(4): 438–446.

Arezou, K., Mahmood, A. S., Maryam, A. L. and Ali, E. 2018. Nanoparticles and their antimicrobial properties against pathogens including bacteria, fungi, parasites and viruses. Microbial Pathogenesis 123: 505–526.

Bugra, D., Aidai, D. K. and Hatice, A. A. 2019. Characterization of recuperating talent of white-rot fungi cells to dye-contaminated soil/water. Chinese Journal of Chemical Engineering 27: 634–638.

Chandran, K., Stacey, L. H. and Soon, Y. 2016. *In vivo* toxicological assessment of biologically synthesized silver nanoparticles in adult Zebrafish (Danio rerio). Journal of Hazardous Materials 301: 480–491.

Cíntia, L. H., Fernando, S. L., Marcela, F. G. G., Meg da, S. F., Sandra, R. G., Elza Iouko, I. et al. 2019. Parameters of the fermentation of soybean flour by *Monascus purpureus* or *Aspergillus oryzae* on the production of bioactive compounds and antioxidant activity. Food Chemistry 271: 274–283.

Constanza, B. L., María, S. J .T., Emilce, V., Marcela, A. F. and María, E. L. 2019. Development of low-cost formulations of plant growth-promoting bacteria to be used as inoculants in beneficial agricultural technologies. Microbiological Research 219: 12–25.

Dapurkar, D. and Telang, M. 2017. A patent landscape on application of microorganisms in construction industry. World Journal of Microbiology and Biotechnology 33(7): 13–18.

De, D., Ananda Raja, R., Ghoshal, T. K., Mukherjee, S. and Vijayan, K. K. 2018. Evaluation of growth, feed utilization efficiency and immune parameters in tiger shrimp (Penaeus monodon) fed diets supplemented with or diet fermented with gut bacterium *Bacillus* sp. DDKRC1 isolated from gut of Asian seabass (Lates calcarifer). Aquacultural Resaerch 49: 2147–2155.

Deniz, U., Jasmin, P., Ilona, L., Martin, H. E., Annette, R. et al. 2019. Drivers of entomopathogenic fungi presence in organic and conventional vineyard soils. Applied Soil Ecology 133: 89–97.

Eduardo, J. F., Angels, R., Nerea, R., Eugenia, Z., Carlos, I. and Catalina F. D. 2014. Nanoparticles as a novel delivery system for vitamin C administration in aquaculture. Aquaculture 432: 426–433.

Egert, M. and Simmering, R. 2016. The microbiota of the human skin. Advances in Experimental Medicine and Biology 902: 61–81.

Fadhile, A., Hassimi, A. H., Mushrifah, I., Siti Rozaimah, S. A., Nurina, A. and El Mubarak, M. T. 2017. Biosurfactant production by the hydrocarbon-degrading bacteria (HDB) *Serratia marcescens*: Optimization using central composite design (CCD). Asian Journal of Industrial Engineering Chemistry 47: 272–280.

Fenghua, W., Tongtong, Z., Lusheng, Z., Xiuguo, W., Jun, W., Jinhua, W. et al. 2019. Effects of successive metalaxyl application on soil microorganisms and the residue dynamics. Ecology Indicators 103: 194–201.

François, P. D. and Willem, M. V. 2019. Biotechnology of health-promoting bacteria. Biotechnological Advances 1–12.

Gharieb, S. El-S., Farag, M. M. and Ahmed, I. El-B. 2018. One-pot green synthesis of magnesium oxide nanoparticles using *Penicillium chrysogenum* melanin pigment and gamma rays with antimicrobial activity against multidrug-resistant microbes. Advanced Powder Technology 29(2018): 2616–2625.

Goda, A. M., Omar, E. A., Srour, T. M., Kotiet, A. M., El-Haroun, E. and Davies, S. J. 2018. Effect of diets supplemented with feed additives on growth, feed utilization, survival, body composition and intestinal bacterial load of early weaning European seabass, *Dicentrarchus labrax* post-larvae. Aquacultural International 26: 169–183.

Gonzalez Pereyra, C. L., Martínez, A. P., Petroselli, G., Erra Balsells, R. and Cavaglieri, L. R. 2018. Antifungal and aflatoxin-reducing activity of extracellular compounds produced by soil Bacillus strains with potential application in agriculture. Food Control 85: 392–399.

Hafiza, F., Faizah, U., Amna, T., Viqar, S., Madeeha, A. and Viqar, U. A. 2019. Evaluation of antimicrobial potential of endophytic fungi associated with healthy plants and characterization of compounds produced by endophytic *Cephalosporium* and *Fusarium solani*. Biocatalysis and Agricultural Biotechnology 18: 101043.

He, H., Deng, D. and Hwang, H. M. 2019. The current application of nanotechnology in food and agriculture. Journal of Food and Drug Analysis 27: 1–21.

Held, N. A., McIlvin, M. R., Moran, D. M., Laub, M. T. and Saito, M. A. 2019. Unique patterns and biogeochemical relevance of two-component sensing in marine bacteria. mSystems 4(1): 1–10.

Hulkoti, N. I. and Taranath, T. C. 2014. Biosynthesis of nanoparticles using microbes—A review. Colloidal surfaces B: Biointerfaces 121: 474–483.

James, R., Kamila, G., Joanna, V., Richard, R., Lubomira, T. and Svetlana, M. 2017. Application of Cu-FAU nanozeolites for decontamination of surfaces soiled with the ESKAPE pathogens. Microporus and Mesoporous Materials 253: 233–238.

Jose, P. A. and Jha, B. 2016. New dimensions of research on actinobacteria: Quest for next generation antibiotics. Froniers in Microbiology 7: 1295.

Jung, J., Yoo, J. E., Choe, Y. H., Park, S. C., Lee, H. J. and Lee, H. J. 2019. Cleaved cochlin sequesters *Pseudomonas aeruginosa* and activates innate immunity in the inner ear. Cell Host. Microbes 25(4): 513–525.

Kallscheuer, N., Classen, T., Drepper, T. and Marienhagen, J. 2019. Production of plant metabolites with applications in the food industry using engineered microorganisms. Current Opinion in Biotechnology 56: 7–17.

Kannan, M., Samuthirapandian, R., Thirunavukkarasu, M., Venkatachalam, U., Ramachandran, C., Palaniappan, S. et al. 2019. Potential uses of fungal polysaccharides as immunostimulants in fish and shrimp aquaculture: A review. Aquaculture 500: 250–263.

Kashyap, P. C., Chia, N., Nelson, H., Segal, E. and Elinav, E. 2017. Microbiome at the frontier of personalized medicine. Mayo Clinic Proceedings 92(12): 1855–1864.

Khorasani, S., Danaei, M. and Mozafari, M. R. 2018. Nanoliposome technology for the food and nutraceutical industries. Trends in Food Science and Technology 79: 106–115.

Koehler, A. M. and Shew, H. D. 2019. Effects of fungicide applications on root-infecting microorganisms and overwintering survival of perennial stevia. Crop Protection 120: 13–20.

Kuebutornyea, F. K. A., Abarikea, E. D. and Lu, Y. 2019. A review on the application of Bacillus as probiotics in aquaculture. Fish and Shell fish Immunology 87: 820–828.

Li, H. L., Li, X. M., Mándi, A., Antus, S., Li, X. and Zhang, P. 2017. Characterization of cladosporols from the marine algal-derived endophytic fungus *Cladosporium cladosporioides* EN-399 and configurational revision of the previously reported cladosporol derivatives. Journal of Organic Chemistry 82(19): 9946–9954.

Li, Y., Jiang, X., Hao, J., Zhang, Y. and Huang, R. 2019. Tea polyphenols: The application in oral microorganism infectious diseases control. Archives of Oral Biology 102: 74–82.

Liu, H., Shi, C., Wu, T., Jia, Q., Zhao, J. and Wangf, X. 2017. Isolation and characterization of methanethiol-producing bacteria from agricultural soils. Pedosphere 27(6): 1083–1091.

Liu, L., Li, H., Xu, R. H. and Li, P. L. 2017. Expolysaccharides from *Bifidobacterium animalis* RH activates RAW 264.7 macrophages through toll-like receptor 4. Food and Agricultural Immunology 28(1): 149–161.

London, L. E. E., Chaurin, V., Auty, M. A. E., Fenelon, M. A., Fitzgerald, G. F., Ross, R. P. et al. 2015. Use of Lactobacillus mucosae DPC 6426, an exopolysaccharide-producing strain, positively influences the techno-functional properties of yoghurt. Internat. Dairy Journal 40: 33–38.

Lu, Z. Q., Jin, M. L., Huang, M., Wang, Y. M. and Wang, Y. Z. 2013. Bioactivity of selenium-enriched exopolysaccharides produced by Enterobacter cloacae Z0206 in broilers. Carbohydrate Polymers 96(1): 131–136.

Lynch, K. M., Coffey, A. and Arendt, E. K. 2017. Exopolysaccharide producing lactic acid bacteria: Their techno-functional role and potential application in gluten-free bread products. Food Research International 110: 52–61.

Madhurankhi, G. and Suresh, D. 2019. Biosurfactant production by a rhizosphere bacteria *Bacillus altitudinis* MS16 and its promising emulsification and antifungal activity. Colloids Surfaces B: Biointerfaces 178: 285–296.

Maeda, H., Zhu, X., Suzuki, S., Suzuki, K. and Kitamura, S. 2004. Structural characterization and biological activities of an exopolysaccharide kefiran produced by *Lactobacillus kefiranofaciens* WT-2B. Journal of Agricultural and Food Chemistry 52(17): 5533–5538.

Magarvey, N. A., Keller, J. M., Bernan, V., Dworkin, M. and Sherman, D. H. 2004. Isolation and characterization of novel marine-derived actinobacteria taxa rich in bioactive metabolites. Applied Environment and Microbiology 70(12): 7520–9.

Manjunath, H. M. and Chandrashekhar, G. J. 2019. Characterization, antioxidant and antimicrobial activity of silver nanoparticles synthesized using marine endophytic fungus—*Cladosporium cladosporioides*. Process Biochemistry 2019: 1–6.

Marienhagen, J. and Bott, M. 2013. Metabolic engineering of microorganisms for the synthesis of plant natural products. Journal of Biotechnology 163(2): 166–78.

Mary-Anne, L., Falko, M., Meng, H. L., Gavan, M. and Daniel, V. M. 2019. Matthias Leopold. *Bacillus subtilis* and surfactant amendments for the breakdown of soil water repellency in a sandy soil. Geoderma 344: 108–118.

Masand, M., Jose, P. A., Menghani, E. and Jebakumar, S. R. 2015. Continuing hunt for endophytic actinobacteria as a source of novel biologically active metabolites. World Journal of Microbiology and Biotechnology 31(12): 1863–75.

Matsumoto, A. and Takahashi, Y. 2017. Endophytic actinobacteria: promising source of novel bioactive compounds. Journal of Antibiotics (Tokyo) 70(5): 514–519.

Mercedes, G. S., Tereza, S., Inmaculada, G. R., Jirina, S. and Pavel, T. 2017. Risk element immobilization/stabilization potential of fungal-transformed dry olive residue and arbuscular mycorrhizal fungi application in contaminated soils. Journal of Environmental Management 201: 110–119.

Milagre, A. P., Daylin, R. R., Edson, R. V., Adriana, F. S. and Marcos, A. C. L. 2019. Conversion of renewable substrates for biosurfactant production by *Rhizopus arrhizus* UCP 1607 and enhancing the removal of diesel oil from marine soil. Electronic Journal of Biotechnology 38: 40–48.

Misbahud, D., Rubina, N., Muhammad, S. A., Faisal, H. K., Asad, K., Munib, A. et al. 2019. Production of nitrogen fixing *Azotobacter* (SR-4) and phosphorus solubilizing *Aspergillus niger* and their evaluation on *Lagenaria siceraria* and *Abelmoschus esculentus*. Biotechnology Reports 20: 1–5.

Mohamed, S., Mona, S., Magdy, El-M. and Mansour, El-M. 2016. Recent progress in applications of nanoparticles in fish medicine: A review. Nanomedicine: Nanotechnology, Biology, and Medicine 12: 701–710.

Morena, G., Chiara, G., Vincenzo, L., Jeannette, L., Llaria, D. and Laura, P. 2016. The impact of mycorrhizal fungi on Sangiovese red wine production: Phenolic compounds and antioxidant properties. LWT—Food Science and Technology 72: 310–316.

Motahhareh, A., Hassan, E. and Hossein, A. A. 2018. Characterization of rhizosphere and endophytic bacteria from roots of maize (*Zea mays* L.) plant irrigated with wastewater with biotechnological potential in agriculture. Biotechnology Reports 20: 1–8.

Nacide, K., Sahin, C., Fatih, D. K., Husniye, A. S. and Cengiz, D. 2019. How titanium dioxide and zinc oxide nanoparticles do affect soil microorganism activity? European Journal of Soil Biology 91: 18–24.

Nadia, B., Montserrat, P., Saoulajan, C., Naima, El M., Angeles, M. and David, M. G. 2018. Isolation and characterization of halophilic bacteria producing exopolymers with emulsifying and antioxidant activities. Biocatalysis and Agricultural Biotechnology 16: 631–637.

Niladri, C., Sourabh, D., Vasvi, C., Anuradha, S., Quaiser, S., Ameer, A. et al. 2018. Bio-inspired nanomaterials in agriculture and food: Current status, foreseen applications and challenges. Microbial Pathogenesis 123: 196–200.

Nivien, A. N., Elhagag, A. H., Mohamed, H. A. and Magdy, M. K. B. 2018. Effectiveness of eco-friendly arbuscular mycorrhizal fungi biofertilizer and bacterial feather hydrolysate in promoting growth of *Vicia faba* in sandy soil. Biocatalysis and Agricultural Biotechnology 16: 140–147.

Nur Sazwani, D., Abd Rahman. J. D., Mohamad, A. R., Zaheda, M. A., Nor, Z. O. and Mohamad, R. 2019. *Paenibacillus polymyxa* bioactive compounds for agricultural and biotechnological applications. Biocatalysis and Agricultural Biotechnology 18: 101092.

Olushola, M. A., Ekundayo, O. A. and Kudjo, D. 2019. Arbuscular mycorrhizal fungi and exogenous glutathione mitigate coal fly ash (CFA)-induced phytotoxicity in CFA-contaminated soil. Journal of Environmental Management 237: 449–456.

Parada, J., Rubilar, O., Sousa, D. Z., Martínez, M., Fernández-Baldo, M. A. and Tortella, G. R. 2019. Short term changes in the abundance of nitrifying microorganisms in a soil-plant system simultaneously exposed to copper nanoparticles and atrazine. Science of the Total Environments 670: 1068–1074.

Paula, R. T., Per Bovbjerg, P. and Lars-Flemming, P. 2017. Bacterial activity dynamics in the water phase during start-up of recirculating aquaculture systems. Aquacultural Engineering 78: 24–31.

Per, B. P., Mathis von, A., Paulo, F., Christopher, N., Lars-Flemming, P. and Johanne, D. 2017. Particle surface area and bacterial activity in recirculating aquaculture systems. Aquaculture Engineering 78: 18–23.

Pessôa, M. G., Vespermann, K. A. C., Paulino, B. N., Barcelos, M. C. S. and Pastore, G. M. 2019.Newly isolated microorganisms with potential application in biotechnology. Biotechnology Advances 37(2): 319–339.

Piotr, P., Karolina, K., Janusz, B., Anna, S. O., Anna, S. and Szymon, Z. 2019. Associations between root-inhabiting fungi and 40 species of medicinal plants with potential applications in the pharmaceutical and biotechnological industries. Applied Soil and Ecology 137: 69–77.

Rajendran, V., Arumugam, S. and Subramaniam, G. 2018. Extracellular biosynthesis of silver nanoparticles using *Streptomyces griseoplanus* SAI-25 and its antifungal activity against *Macrophomina phaseolina*, the charcoal rot pathogen of sorghum. Biocatalysis and Agricultural Biotechnology 14: 166–171.

Rajivgandhi, G., Ramachandran, G., Maruthupandy, M., Vaseeharanc, B. and Manoharan, N. 2018. Molecular identification and structural characterization of marine endophytic actinobacteria *Nocardiopsis* sp. GRG 2 (KT 235641) and its antibacterial efficacy against isolated ESBL producing bacteria. Microbial Pathogenesis 126: 138–148.

Ralstonia solanacearum, T. A., Raja Asad, A. K., Ahmad, A., Hassan, S., Zahid, U. and Mohammad, A. 2019. Biogenic synthesis of iron oxide nanoparticles via *Skimmia laureola* and their antibacterial efficacy against bacterial wilt pathogen. Materials Science and Engineering C 98: 101–108.

Ramachandran, G., Rajivgandhi, G., Maruthupandy, M. and Manoharan, N. 2019. Extraction and partial purification of secondary metabolites from endophytic actinobacteria of marine green algae *Caulerpa racemosa* against multi drug resistant uropathogens. Biocatalysis and Agricultural Biotechnology 17: 750–757.

Ramakrishnan, R., Lakkakula, S., Subramani, P., Periyasamy, R., Arokiam, S. and Gowrishankar, S. 2018. Production of squalene with promising antioxidant properties in callus cultures of *Nilgirianthus ciliates*. Industrial Crops & Products 126: 357–367.

Rashad, F. M., Fathy, H. M., El-Zayat, A. S. and Elghonaim, A. M. 2015. Isolation and characterization of multi functional *Streptomyces* species with antimicrobial, nematicidal and phytohormone activities from marine environments in Egypt. Journal of Microbiological Research 175: 34–47.

Ronald, O., Evelyn, A., Grant, B., Friederike, H., Joachim, R. and Gabriela, S. 2001. Sponge-microbe associations and their importance for sponge bioprocess engineering. Hydrobiologia 461: 55–62.

Rong, X., Kai, Z., Pu, L., Huawen, H., Shuai, Z., Apurva, K. et al. 2018. Review lignin depolymerization and utilization by bacteria. Bioresources Technology 269: 557–566.

Sanan, N., Xiumei, L., Lixia, Z., Philip, C. B. and Fei, W. 2018. Fungal communities and functions response to long-term fertilization in paddy soils. Applied Soil and Ecology 130: 251–258.

Saurav, K. and Kannabiran, K. 2012. Cytotoxicity and antioxidant activity of 5-(2,4-dimethylbenzyl) pyrrolidin-2-one extracted from marine *Streptomyces* VITSVK5 spp. Saudi. Journal of Biological Sciences 19: 81–86.

Schreier, H. J., McDonald, R., Marsic-Lucicd, J., GavriloviCe, A., Pecarevica, M. and Jug-Dujakovic, J. 2019. Bacterial community analysis of marine recirculating aquaculture system bioreactors for complete nitrogen removal established from a commercial inoculum Marina Brailoa. Aquaculture 503: 198–206.

Shinde, V. L., Suneel, V. and Shenoy, B. D. 2017. Diversity of bacteria and fungi associated with tarballs: Recent developments and future prospects. Marine Pollution Bulletin 117(1-2): 28–33.

Stout, M. J., Zhou, Y., Wylie, K. M., Tarr, P. I., Macones, G. A. and Tuuli, M. G. 2017. Early pregnancy vaginal microbiome trends and preterm birth. American Journal of Obstetrics and Gynecology 7(3): 1–356.

Subramani, R. and Aalbersberg, W. 2012. Marine actinobacteria: an ongoing source of novel bioactive metabolites. Microbiological Research 67(10): 571–80.

Thea, K., Megan, J. O. M. and Lesley, L. D. 2018. Nanotechnology in the food sector and potential applications for the poultry industry. Trends in Food Sciences and Technology 72: 62–73.

Thiago, C., Johanna, M., Jurg, E., Jorgen, E., Nicolai, V. M. and Rafael de, A. M. 2016. Persistence of Brazilian isolates of the entomopathogenic fungi *Metarhizium anisopliae* and *M. robertsii* in strawberry crop soil after soil drench application. Agricultural Ecosystem and Environment 233: 361–369.

Thorsten, S., Larissa, F., Dominik, F., Tristan, C., Catarina, I. M. M. and Jan, P. 2018. Environmental DNA metabarcoding of benthic bacterial communities indicates the benthic footprint of salmon aquaculture. Marine Pollution Bulletin 127: 139–149.

Tohren, C. G. K. and Keith, A. S. 2019. The effect of nanoparticles on soil and rhizosphere bacteria and plant growth in lettuce seedlings. Chemosphere 221: 703–707.

Vitorino, L. C. and Bessa, L. A. 2017. Technological microbiology: Development and applications. Frontiers in Microbiology 8: 827.

Wei, Q., Wang, X., Sun, D. W. and Pu, H. 2019. Rapid detection and control of psychrotrophic microorganisms in cold storage foods: A review. Trends in Food Sciences and Technology 86: 453–464.

Wentao, L. J., Ruijun, D., Mao, Y., Mingming, S. and Jinzhong, W. F. 2018. Agricultural Waste to Treasure' e Biochar and eggshell to impede soil antibiotics/antibiotic resistant bacteria (genes) from accumulating in Solanum tuberosum. Environmental Pollution 242: 2088–2095.

Wurong, D., Lina, W., Juan, Z., Wencan, K., Jianwei, Z., Jiexu, Z. et al. 2017. Characterization of antioxidant properties of lactic acid bacteria isolated from spontaneously fermented yak milk in the Tibetan Plateau. Journal of Functional Food 35: 481–488.

Wusirika, R., Radheshyam, Y. and Kefeng, Li. 2019. Plant growth promoting bacteria in agriculture: Two sides of a coin. Applied Soil and Ecology 138: 10–18.

Yang, H. and Wang, Y. 2008. Application of atomic force microscopy on rapid determination of microorganisms for food safety. Journal of Food Sciences 73(8): 44–50.

Yang, N. and Song, F. 2018. Bioprospecting of novel and bioactive compounds from marine actinobacteria isolated from south china sea sediments. Current Microbiology 75(2): 142–149.

Yuanmei, X., Yanlong, C., Fangfang, Y., Lihua, L., Yuanyuan, S., Bianfang, L. et al. 2019. Exopolysaccharides produced by lactic acid bacteria and *Bifidobacteria*: Structures, physiochemical functions and applications in the food industry. Food Hydrocolloids 94: 475–499.

Zhang, C. N., Li, X. F., Xu, W. N., Zhang, D. D., Lu, K. L., Wang, L. N. et al. 2015. Combined effects of dietary fructooligosaccharide and *Bacillus licheniformis* on growth performance, body composition, intestinal enzymes activities and gut histology of triangular bream (Megalobrama terminalis). Aquaculture Nutrition 21: 755–766.

Zibo, L., Junfan, F., Rujun, Z. and Dan, W. 2018. Effects of phenolic acids from ginseng rhizosphere on Saudi Journal of Biological Sciences 25: 1788–1794.

CHAPTER 4

Microbial Cell Factories

Asmaa Missoum

Introduction

Microorganisms, which have been inhibiting the earth for about 3.8 billion years, possess important physiological and functional properties. These enable them to act as minute biological factories to synthesize various highly valuable metabolites (Maheshwari et al. 2010). In the 1980s, the biopharmaceutical sector has flourished thanks to ever-increasing knowledge of cell physiology, heterologous gene expression, and protein production. This has empowered the use of different prokaryotic and eukaryotic cells as living factories. Among these systems, *Escherichia coli* and *Saccharomyces cerevisiae* are the most prevalent as they exhibit high versatility, enabling them to be adaptable to different genetic manipulation methodologies as well as production demands (Sanchez-Garcia et al. 2016). Indeed, *E. coli* has known genomics, grows in high densities with a short doubling time, and requires a simple scale-up procedure that is cost-effective. It is also used to produce 70% of the 94 protein-based anti-cancer medications (Negrete and Shiloach 2017). Since the early 1970s, the production of heterologous recombinant proteins in *E. coli* has been a challenge. Approaches such as change of cultivation medium, alternate protein tags, co-expression of chaperons, and use of codon optimized genes, were employed to optimize protein expression (Krause et al. 2016). Alternatively, *S. cerevisiae* is one of the most extremely studied unicellular eukaryotes, which has been traditionally applied in baking processes, alcohol fermentations, and bioethanol production. Novel industrial applications of this yeast are extended to heterologous expression of proteins, as well as the production of fine chemicals such as amorphadiene and vanillin (Kavšček et al. 2015). The main advantages that makes *S. cerevisiae* a prominent host includes rapid growth, vast array of tools for genetic engineering, and a limited native secondary metabolism to minimize potential interference. Likewise, yeast is classified as fungus and thus is capable in expressing important fungal proteins such as cytochrome P450s, as these enzymes often attach to the endoplasmic reticulum, an absent organelle in prokaryotes (Bond et al. 2016).

Paris-sud université, UFR Sciences, 15 Rue Georges Clemenceau, 91400 Orsay, France.
Email: amissoum93@gmail.com

By definition, a microbial cell factory is a bioengineering approach which utilizes microbial cells as production facilities, where the optimization process heavily relies on metabolic engineering. Such approach was proven useful in many biotechnological applications related to medicine, agriculture, environment, as well as other industries (Sharma and Saharan 2018). However, studying the function of genes and proteins has been and remains the main application of cell factory engineering in basic research, along with generating reagents such as hormones and purified proteins for research. For this, synthetic biological tools play major roles in modifying and introducing new gene functions, in addition to analyzing target cellular functions such as biochemical reactions (Davy et al. 2017).

Traditionally, bioactive compounds have been naturally produced by their host cells, e.g., plant or microorganism, either intracellular or secreted compounds. Yet, with the introduction of DNA engineering, strain improvement as well as expressing biosynthetic pathways in heterologous hosts has become feasible. This has enabled efficient, cost effective, and environmentally friendly production of bioactive molecules (Nielsen 2019). Examples of cell factory applications could include and are not limited to: synthesizing plant natural products (PNPs) such as artemisinin, resveratrol, and carotenoids (Liu et al. 2017), biosynthesis of aromatic chemicals including l-tyrosine and l-tryptophan, and l-phenylalanine (Huccetogullari et al. 2019), and pharmacologically valuable nutraceuticals such as polyunsaturated fatty acids, carotenoids, polyphenols, and non-protein amino acids (Yuan et al. 2019). In order to optimize microbial cell factories using metabolic engineering tools, there are several factors that must be taken into account. First, the biosynthesis of natural products is fundamentally associated with precursors, co-factors, and enzymes in primary metabolism. Therefore, it is necessary to consider these three aspects: choice of cell factory, engineering central carbon metabolism, and improving enzymes in the chemical pathways to avoid proteome constraint issue within the cell (Nielsen 2019). Moreover, to increase the formation of important natural products, several strategies can be employed and are often combined for better effects. These are divided into three categories: direct optimization of pathways (substrates' enrichment, feedback inhibition removal, and pathway over expression), removal of competing activities (de-branching, product degradation, and removal of by-products), and application of global regulation engineering, e.g., carbon repression or other signal transduction pathways (Davy et al. 2017). Other strategies such as Division of Labor (DoL), which uses natural or synthetic microbial consortia, can reduce metabolic burden that is usually caused by distributing limited resources among different tasks. Microbial consortia can divide metabolic load amongst partners and has been improved by strain engineering, since they represent dynamic systems that are complicated to operate (Roell et al. 2019).

Although *E. coli* and *S. cerevisiae* are the most widely researched and preferred platforms for the industrial production of various biological products, other species from fungi, microalgae, and cyanobacteria are currently explored (Khan et al. 2018, Knoot et al. 2018). Because there is not much scientific literature that covers metabolic engineering topics concerning microbial factories other than *E. coli* and *S. cerevisiae*, this chapter is dedicated to various fungal, microalgae, and cyanobacterial cell factories. In this context, key examples of these microbial species, which have secured various places in biotechnological production, are presented. Essential strategies and emerging novel microbial hosts that have contributed to successful development of industrial strains, as well as major challenges in the field of metabolic engineering are also discussed in this chapter.

Fungi

Over 1000 years, humans have been using some species of fungi, which play a critical role in many degradation processes and in fighting plant diseases. In medicine, they are utilized to produce antibiotics, e.g., penicillin from *Penicillium chrysogenum* and cephalosporin from *Cephalosporium acremonium*. Other examples concerning current applications of filamentous fungi include the production of cellulase by *Aspergillus niger*, in addition to Kojic acid and tempeh (one of the oldest fermented soy) by *Aspergillus oryzae* (Li et al. 2017). For this reason, the field of fungal biotechnology is expanding as filamentous fungi possess various properties that enable them to be developed as cell factories. These include simple and rapid growth on inexpensive media, as well as capacity in expressing and secreting heterologous proteins. Moreover, genetic transformation and metabolic engineering techniques are developed by researchers in an attempt to genetically modify fungal strains, thus improving target functions (Bischof et al. 2016, Tong et al. 2019).

Approaches for the production of enzymes and secondary metabolites

Genome-scale metabolic models (GEMs) are important tools that assist in designing metabolic engineering strategies for optimization. These are mathematical representations of metabolisms in an organism, which provide better understanding of secondary metabolite production. For instance, a reconstructed GEM of penicillin over-producing strain, *Penicillium chrysogenum* Wisconsin 54-1255, suggested that increasing NADPH availability and modifying precursor supplying pathways could maximize penicillin production. Moreover, GEM of *Aspergillus nidulans* was used to calculate the metabolic fluxes under the over expression of xylulose-5-phosphate phosphoketolases (XPKs). The analysis implied that inducing XPKs increased the carbon flux towards acetyl-CoA, the precursor for polyketide synthases (PKS) biosynthesis (Nielsen and Nielsen 2017).

Salo and his co-workers compared secondary metabolite biosynthesis in penicillin over-producer *Penicillium chrysogenum* DS17690, with strain DS68530 that lost its penicillin producing ability. Using genomic analysis, they found out that the production of other non-ribosomal peptides (NRPs) like chrysogines and roquefortines/meleagrin were increased in strain DS68530. It was explained that a re-direction of nitrogen metabolism toward other NRPs has taken place (Salo et al. 2015). Moreover, another very recent genomic study identified a novel 13-membered calbistrin producing gene cluster (*calA* to *calM*) in *Penicillium decumbens*. This metabolite is known to have bioactivities against leukemia cells. It was also observed that the deletion of polyketide synthase (*calA*), major facilitator pump (*calB*), and a transcription factor (*calC*) encoding genes using CRISPR/Cas9 technology resulted in no production of calbistrin. This strongly proves the involvement of previously mentioned genes in the biosynthesis of calbistrins (Grijseels et al. 2018).

Another technique to enhance the secondary metabolite production is employing promoter exchange to achieve an inducible pathway. The promoter *acvA* in *Aspergillus nidulans*, which expresses ACV synthetase (the rate limiting enzyme of the penicillin pathway), was replaced by an inducible alcohol dehydrogenase 1 (alcAp) promoter and resulted in a 30-fold increase in penicillin yields (Nielsen and Nielsen 2017). Domain swapping or replacement by synthetic versions in PKS and NRPS engineering is an additional strategy. Recently, swapping two PKS-NRPS natural hybrids CcsA from *Aspergillus clavatus* (produces cytochalasin E) and Syn2 from *Magnaporthe oryzae* leads to production of novel hybrid products such as niduporthin and niduclavin along with niduchimaeralin A and B. These constructed functional cross-species PKS-NRPS were expressed in *Aspergillus nidulans* (Guzmán-Chávez et al. 2018a).

Intriguingly, deleting highly expressed NRPS gene *Pc21g12630* (*chyA*) resulted in decreased biosynthesis of chrysogine and 13 other related compounds in *Penicillium chrysogenum*. In *chyA*-containing gene cluster, each gene was individually deleted to assess its function and it was concluded that the biosynthetic pathway was highly branched and complex, with few enzymes involved in many steps of the biosynthesis (Viggiano et al. 2018). In another research, high levels of sorbicillinoid production were restored by repairing a mutation in polyketide synthase SorA and SorB enzymes in *Penicillium chrysogenum* DS68530. These are required to generate sorbicillin and dihydrosorbicillin, which are the key intermediates in the sorbicillinoid biosynthesis pathway. Additionally, authors discovered through gene deletion and metabolite profiling, that this pathway is regulated by *sorR1* and *sorR2* transcription factors while sorbicillinoids act as autoinducers (Guzmán-Chávez et al. 2017). Gene silencing of four genes (prx1, prx2/ari1, prx3 and prx4) from a cluster of PR-toxin biosynthesis, a potent mycotoxin, reduced its production by 65–75% in *Penicillium roqueforti.* However, these four silenced mutants overproduced huge amounts of mycophenolic acid, an antitumor compound synthesized by an unrelated pathway, signifying a cross-talk of mycophenolic acid and PR-toxin production (Hidalgo et al. 2014).

Transformation methods are also utilized to genetically modify fungal strains by inserting exogenous DNA to obtain enhanced features. However, due to their immense diversity, different methods might be required for different species. Hence, transformation protocols must be optimized for each strain. Protoplast-mediated transformation (PMT), electroporation, *Agrobacterium*-mediated transformation, biolistic method, and shock-wave-mediated transformation are among the most widely used methods (Li et al. 2017). Some of these methods served for the heterologous expression of cytochrome P450 (CYP) monooxygenases, which are ubiquitous enzymes that are versatile and promising catalysts, represented in Fig. 1. For instance, two CYPs (PbP450-1 and PbP450-2) along with geranylgeranyl diphosphate synthase and terpene synthase were introduced in *Aspergillus oryzae* for the biosynthesis of diterpene aphidicolin. Similarly, controlled expression of two CYPs (*tenA* and *tenB* from *Beauveria bassiana*) with *amyB* inducible promoter enhanced tenellin productivity (243 mg/L) by 5-fold in *A. oryzae* compared to

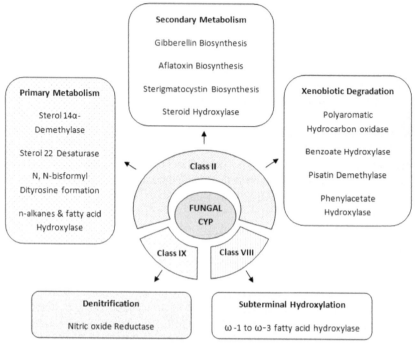

Fig. 1. Representative diagram of functional classification of diverse fungal CYP enzymes (Durairaj et al. 2016).

the wild-type. Moreover, the co-expression of *tenA* and *tenB* genes coupled with hybrid polyketide synthetases in *A. oryzae* facilitated bassian in synthesis (Durairaj et al. 2016).

Applications of Penicillium chrysogenum, Trichoderma reesei, and Aspergillus niger

Since the discovery of penicillin by Alexander Fleming, the genus *Penicillium* has been extensively studied for its ability to produce a large range of natural products. Among more than 350 species in this genus, *Penicillium chrysogenum*, which was renamed as *P. rubens*, is the most studied member and is known to produce chrysogine, fungisporin, siderophores, ω-hydroxyemodin, roquefortines, and penitric acid. To improve penicillin titers, the strain was subjected to a CSI (classical strain improvement) program. Although it resulted in higher penicillin productivity, it also reduced the synthesis of various BGC (biosynthetic gene clusters) encoded natural products. Consequently, several approaches have been used to resolve this issue (Guzmán-Chávez et al. 2018a).

One of the most powerful tools for engineering *P. chrysogenum* is the CRISPR/ Cas9 based system. It was recently developed for genome modifications and the study showed that deleting full gene clusters is possible with minimal cloning efforts. Along with enhanced homologous recombination, this opens the possibilities to engineer new synthetic pathways in *P. chrysogenum* and re-factoring it as platform organism

(Pohl et al. 2018). In a novel study, deleting *Pc21 g14570* gene coding for class 2 histone deacetylase (HdaA) ortholog induced the expression of certain polyketide synthase (PKS) and non-ribosomal peptide synthetase (NRPS) encoding genes. This in turn resulted in lower levels of chrysogine, the overproduction of sorbicillinoids, and the detection of a new unknown compound under these conditions. Such findings showed that HdaA mediates the transcriptional crosstalk between sorbicillinoids biosynthesis and other BGCs (Guzman-Chavez et al. 2018b).

Another interesting fungus, *Trichoderma reesei*, was isolated from the Solomon Islands around 70 years ago. This was owing to its ability to degrade cellulose containing fabrics, a trait that depends on its secreted cellulases, which is nowadays exploited by numerous industries. Most importantly, *T. reesei* enzymes are used to saccharify lignocellulose obtained from renewable plant biomass to generate biobased fuels and other chemicals (Bischof et al. 2016). Other produced metabolites by different engineered strains of *T. reesei* are presented in Table 1 below.

It is well known that the transcription of cellulase and xylanase strictly relies on the $Zn(II)_2Cys_6$ type transcriptional activator XYR1. The over expression of *xyr1* was proven to abolish the catabolite repression of cellulases in the presence of glucose and augment their expression levels. Moreover, deregulation of *xyr1* mediated by *tcu1* (copper transporter encoding gene) promoter resulted in full expression of cellulases on the non-inducing carbon sources and in the absence of copper (Lv et al.

Table 1. Summary of some genetically engineered *Trichoderma reesei* strains and their interesting metabolites (Bischof et al. 2016).

T. reesei strain(s)	Substance generated	Substrate used	Maximum titer	Genetic modification
RUT-C30, QM6a	Erythritol	Wheat straw	5 mg l^{-1}	Erythrose reductase expression using the *pki1* and *bxl1* promoters
QM9414	Ethylene	Wheat straw	4.012 nl h^{-1} l^{-1}	Ethylene forming enzyme from *P. syringae* expressed using *cel7a, pgk1* and *gpdA* (*A. nidulans*) promoters
TL124	Ethylene	Wheat straw, cellulose	1.06 µl h^{-1} g^{-1} (dry weight)	Ethylene forming enzyme from *Pseudomonas syringae* expressed using *cel7a* promoter
QM9414	N-acetyl-neuraminic acid	Chitin	13 µg g^{-1} (mycelium)	Expression of codon optimized N-acetyl glucosamine-2-epimerase (*Anabaena* sp.) and N-acetyl neuramininc acid synthase (*Campylobacter jejuni*) using *pki1* and *xyn1* promoters
QM9414	Xylitol	Barley straw	13.2 g l^{-1}	Deletion of xylitol dehydrogenase and l-arabinitol-4-dehydrogenase
QM9414	Xylitol	Xylose and glucose	3.7 g l^{-1}	Over expression of d-xylose reductase, deletion of xylitol dehydrogenase, and knockdown of xylulokinase (antisense RNA)

2015). Besides *xyr1*, three transcription factors ACE1, ACE2 and ACE3 also regulate xylanase and cellulase expression in *T. reesei*. The deletion of ACE1, which is a C_2H_2 zinc finger repressor, was found to improve the production of both xylanases and cellulases (Häkkinen et al. 2014).

Another key player in cellulase and hemicellulase production is the manipulation of Velvet-LaeA/LAE1 complex. These are group of genes that participate in chromatin modification. LaeA, which has characteristics of an *S*-adenosyl-l-methionine arginine protein methyltransferase, seems to function by interacting with transcription factors of the Velvet protein complex. In *T. reesei*, the over expression of both *vel1* and *lae1* has been shown to boost cellulase expression and secretion. In contrast, deleting *vel1* completely impaired the expression of xylanases, cellulases, as well as the cellulase regulator *xyr1* on lactose as an inducing carbon source (Druzhinina and Kubicek 2017).

Aspergillus niger is an important cell factory for the industrial production of citric acid, which is the world's most consumed organic acid as it is widely used in food, beverage, and pharmaceutical industries. This fungus also has powerful degrading enzymes that hydrolyze many polymeric substrates, as well as inherent physiological characters for industrial fermentation. Despite this, a gap still exists between the practical and the theoretical yield of citric acid (Tong et al. 2019). Thus, current metabolic engineering efforts are employed to boost citric acid accumulation in *A. niger*.

To improve the efficiency of citric acid production, residual sugar that remains at the end of the fermentation process is reduced. Deletion of α-glucosidases encoding for *agdA* gene decreased the iso-maltose concentration, a main component of residual sugar. In addition to this, the over-expression of glucoamylase *glaA* decreased residual sugar by 88.2% and increased citric acid production by 16.9%, generating up to 185.7 g/L (Wang et al. 2016). Furthermore, the deletion of two cytosolic ACL subunits in *A. niger* strain ATCC1015 resulted in declining citric acid production with diminished conidial germination and cell growth. Yet, the overexpression exhibited opposite effects, suggesting that ACL is valuable for citric acid accumulation and may be involved in the ATP futile cycle (Yin et al. 2017).

Interestingly, citric acid and ATP are inhibitors of PFK, which is a controlling step for glycolysis metabolic flux. It was reported that a team of researchers designed a PFK1 fragment (*mt-pfkA10*) with T89D single site mutation to avoid the phosphorylation requirement. Its over expression in *A. niger* strain A158 exhibited a citric acid production that is 70% higher than the control strain, reaching up to 120 g/L at 300 h (Tong et al. 2019).

In submerged fermentation, the alteration of *A. niger* morphology plays a vital role in citric acid production as it results in the aggregation of mycelia pellets, which is the ideal morphology. A recent study showed that silencing of the chitin synthase gene (*chsC*) by RNA interference enhanced citric acid accumulation by 42.6%. Since chitin is a crucial component of the fungal cell wall, regulating its synthesis could have profound effects on the fungal morphology during submerged culture (Sun et al. 2018).

Cyanobacteria

Cyanobacteria, which originated around 2.6 to 3.5 billion years ago, are known as the oldest photosynthetic microorganisms on earth. Unlike heterotrophic bacteria, they don't require complex growth media and carbon sources for growth. As a consequence, the production of biochemicals and biofuels by cyanobacteria from solar energy and CO_2 (a greenhouse gas) offers an eco-friendly path. Moreover, cyanobacteria convert around 3–9% of the solar energy into biomass when compared to 0.25–3% achieved by plants, making them more superior (Knoot et al. 2018). They are also characterized by relatively fast growth, demand less land for cultivation, reducing competition with crops determined for human consumption, and their left-over biomass can be converted to organic fertilizers or used as animal feed. In terms of strain engineering, cyanobacteria can be easily manipulated due to their prokaryotic, simple genetic background (Calero and Nikel 2018). The main genetic engineering strategies that have been used for improving the production of cyanochemicals are summarized in Fig. 2. These include enhancing CO_2 fixation efficiency, distributing endogenous carbon flux, redox balance and product conversion efficiency (Zhou et al. 2016).

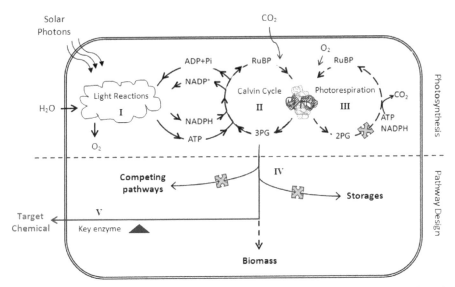

Fig. 2. Genetic engineering strategies that facilitate cyano-factories development. (*I*) Improving of photosynthesis with enhancement of light reactions, (*II*) optimizing RuBisCO to increase Calvin cycle efficiency, and (*III*) disrupting photorespiration pathway to decrease photorespiration. Other targets include (*IV*) blocking competing and storages pathways and (*V*) increasing expression level of key enzymes, considering co-factor balance using NADPH-dependent enzymes (Zhou et al. 2016).

Biofuels production in cyanobacteria

Although some species produce small amounts of ethanol as a byproduct of natural fermentation, engineered *Synechocystis* sp. was reported to produce up to 5500 mg/L. This was achieved through overexpression of endogenous alcohol dehydrogenase as well as disrupting the biosynthetic pathway of polyhydroxyalkanoate (Al-Haj et al. 2016). However, introducing two copies of the *adc* gene into *Synechocystis* sp. PCC 6803 enhanced ethanol production from 1 g/L to 5.5 g/L (Zhou et al. 2016). Isobutanol, a better substitute to ethanol, was photo-synthetically produced (450 mg/L) by introducing an artificial biosynthesis pathway into *Synechococcus elongatus* PCC 7942 (Al-Haj et al. 2016).

Hydrogen is also produced as a byproduct of nitrogen fixation by many genera including *Anabaena*, *Calothrix*, and *Oscillatoria*. Increased production of hydrogen is possible by blocking competing pathways with hydrogenases, e.g., *ldhA* gene that consumes NADH in *Synechococcus* sp. PCC 7002 lactate production, or by heterologous expression of external hydrogenases from other strains, e.g., *Clostridium acetobutylicum*. Another metabolic engineering strategy includes redirecting glycogen catabolism and increasing intracellular NADPH concentrations. For example, deleting glyceraldehyde-3-phosphate (*gap*1) gene and overexpressing NAD^+-glyceraldehyde-3-phosphate (GAPDH-1) has enhanced hydrogen yield by 2.3-fold and 3-fold, respectively, in *Synechococcus* sp. PCC 7002 (Kumaraswamy et al. 2019).

The ability for cyanobacteria to synthesize alkanes or alkenes is owing to two essential enzymes, aldehyde decarbonylase and acyl-acyl carrier protein reductase, which use intermediates of fatty acid metabolism. It was shown that the overexpression or the heterologous expression (from *S. elongatus* PCC 7942 to *Synechococcus* sp. PCC 7002) of these two enzymes significantly increased the yield of alkanes and alkenes (Knoot and Pakrasi 2019). In addition, overexpressing acyl-ACP reductase, an aldehyde forming enzyme in *Synechocystis* sp. PCC 6803, was also found to be useful in enhancing fatty acids and alkane production (Eungrasamee et al. 2019).

Other potential biofuels such as isoprene can be synthesized by transforming *Synechocystis* sp. with isoprene synthase (*IspS*) genes from *Pueraria montana,* using heterologous expression. The co-introduction of seven mevalonic acid pathway genes was also found to increase isoprene production yield to nearly 2.5-fold (Xue and He 2015). Additionally, alternative energy sources such as microbial fuel cell (MFC) can also be used to recover energy in the form of electricity. These convert solar or chemical energy to electrical energy via microorganisms as catalysts. Different cyanobacterial species have been evaluated as the electricigens in photosynthetic MFCs (PMFCs); bioelectrochemical systems utilize light as source of power to generate electricity by light-driven oxidation of water (Ng et al. 2017). For example, the maximum power output of *Synechocystis* PCC-6803 PMFC was 72.3 mW/m^2 compared to 6.5 mW/m^2 from *Spirulina platensis* PMFC. Although efficiency of electricity generation of the PMFCs is still low, in progress research is improving the power generation ability by externally adding feedstocks such as 1,4-benzoquinone as electron mediator in *Nostoc* sp. ATCC 27893 PMFC, which has generated 100 mW/m^2. Genetically modified current electricigens with excellent electrochemical activities is also considered (Cao et al. 2019).

Cyanobacteria as sources of bioactive and value-added compounds

Cyanobacteria excrete bioactive compounds as allelochemicals, which have been found to have important bioactivities such as antimicrobial, anti-inflammatory, anticancer, and so forth. These include lipopeptides (40%), amino acids (5.6%), macrolides (4.2%), and amides (9%) (Table 2). Other commercially valuable products such as enzymes (protease, amylase, and phosphatases), pigments (Carotenoids and phycobiliproteins), and vitamins (vitamins B, B12, and E) are also produced at large-scale (Lau et al. 2015).

Polyhydroxyalkanoate (PHA), which is a type of biodegradable polymer, is also reported to be naturally synthesized by various cyanobacteria such as *Aphanothece* sp., *Oscillatoria limosa*, and *Synechococcus* sp. MA19. PHA accumulation is enhanced by several strategies including the introduction of acetoacetyl-CoA synthase genes in *Synechocystis* sp., as well as plasmid transformation with *Cupriavidus necator* PHA biosynthetic genes, where *recA* complementation is used as selection pressure for plasmid stability. 3-Hydroxybutyrate, a PHB metabolic precursor, has also been

Table 2. Important molecules produced by cyanobacteria and their bioactivities (Lau et al. 2015).

Bioactive/value compound	Producer cyanobacterium	Activity/application
Cyanobacterin	*Scytonema hofmanni*	Antialgal
Enediyne-containing Photosystem II Inhibitor	*Fischerella muscicola*	
Hapalindoles	*Hapalosiphon* and *Fischerella* spp.	
Nostocyclamides	*Nostoc* sp.	
Noscomin	*Nostoc commune*	Antibacterial
Nostofungicide	*Nostoc commune*	Antifungal
Norharmane	*Nostoc insulare*	
4,4'-Dihydroxybiphenyl	*Nodularia harveyana*	
Cyanovirin-N	*Nostoc ellipsosporum*	Antiviral
Scytovirin N	*Scytonema varium*	
Sulfoglycolipid	*Scytonema* sp.	
Vitamin B12, beta-carotene, thiamine, and riboflavin	*Spirulina* (*Arthrospira*)	Commercial capsules/tablets
Proteases	*Anabaena variablis*	Food processing industries
Alpha-amylases		Starch industries
Phosphatases		Diagnostic markers
Phycobiliproteins		Food colorants, food additives, and supplements for human and animal feeds
Carotenoids		Antioxidant, anticancer, and anti-cardiovascular problems
Polyhydroxyalkanoate (PHA)	*Aphanothece* sp., *Oscillatoria limosa*, *Synechococcus* sp. MA19	Biodegradable polymer

synthesized in *Synechocystis* sp. PCC 6803 in order to reduce the high production costs associated with traditional fermentation approaches (Calero and Nikel 2018).

Production of plant secondary metabolites

Cyanobacteria are suitable to be used as cell factories for biosynthesizing plant secondary metabolites thanks to many properties. These include their amenability to genetic engineering, photosynthetic activity, and their ability to live in harsh environments. A broad variety of enzymes and pathways are involved in the production of plant secondary metabolite. However, it is important to consider that some plant enzymes require post-translational modifications, e.g., glycosylation. Thus, they could be non-functional when expressed in cyanobacteria as they are not equipped with these machineries. To produce ethylene, two ethylene-generating enzymes from *Solanum lycopersicum* (tomato) were inserted into *S. elongatus* PCC 7942, generating a titer of $\sim 3.9 \, \mu g$ ethylene $L^{-1} h^{-1} OD730^{-1}$ (Jindou et al. 2014).

To increase limonene production, an important medicinal chemical and several metabolic strategies were used. These included expressing limonene synthase gene (*LMS*) from *Schizonepeta tenuifolia* in *Synechocystis* sp. PCC 6803 (41 $\mu g L^{-1} day^{-1}$), expressing codon optimized limonene synthase gene from *Mentha spicata* (*mslS*) in *Synechococcus* sp. PCC 7002 (50 $\mu g L^{-1} h^{-1}$), and overexpressing three genes (*crtE, dxs*, and *ipi*) in the 2-C-methyl-d-erythritol-4-phosphate (MEP) pathway (Xue and He 2015). The latter augmented the supply of geranyl pyrophosphate (GPP), a limonene substrate, resulting in a 1.4-fold improved production. Carotenoids are also naturally produced through MEP pathway. It was found that overexpressing 1-deoxy-d-xylulose-5-phosphate synthase (DXS) in *Synechocystis* sp. PCC 6803 has boosted the carotenoid level by 1.5 times higher (Knoot et al. 2018).

Other vital metabolites such as caffeic acid can be synthesized in *Synechocystis* sp. PCC 6803 by expressing *Arabidopsis* ρ-coumarate 3-hydroxylase, with the addition of ρ-coumaric acid as substrate to medium. A further genetic engineering technique consists of producing ρ-coumaric acid by *Synechocystis* sp. PCC 6803, a mutant lacking a laccase gene and harboring a tyrosine ammonia-lyase (TAL) gene from *Saccharothrix espanaensis* (Xue et al. 2014). It was also reported that mannitol was successfully produced in *Synechococcus* sp. PCC 7002 by inactivating glycogen synthesis (knocking out *glgC* gene), as well as heterologously expressing mannitol-1-phosphatase (*mlp*) from *Eimeria tenella* and mannitol-1-phosphate dehydrogenase (*mtlD*) from *E. coli*. This has resulted in a titer of 0.15 g $L^{-1} day^{-1}$ with increased yield of 3.2-fold (Madsen et al. 2018).

Microalgae

Microalgae are a diverse group of unicellular, photoautotroph, and microscopic algae that have simple growth requirements. Their faster growth rates compared to terrestrial plants offer several advantages for algal biotechnology, as numerous species such as *Chlorella* and *Dunaliella* were exploited for the production of valuable macromolecules (e.g., carotenoids, phycocolloids, vitamins, and fatty acids), and bioactive compounds (e.g., antioxidant, antimicrobial, anti-inflammatory, and antitumor). However, micro algal compounds remain largely unexplored due to a number of economic barriers. Hence, algal biotechnology companies have currently begun to explore and implement the strategies of engineering of metabolic pathways to synthesize increased levels of desirable compounds (García et al. 2017).

Approaches for developing algal cell factories

Unlike cyanobacteria, microalgae are photosynthetic eukaryotic organisms. There are lots of parameters that can be utilized to increase biochemicals' production in microalgae. These include mutagenesis. In this process, chemical or physical mutagens are used to induce mutations at higher rates in a particular organism. Consequently, stable mutants with enhanced traits are selected (Khan et al. 2018). For example, N'-nitro-N-nitrosoguanidine (NTG) is a chemical mutagen that has successfully enhanced carotenoid and astaxanthin content accumulation in *Haematococcus pluvialis*, and ethyl methanesulfonate (EMS) managed to create *Chlamydomonas reinhardtii* mutants with increased lipid production. Other physical mutagens such as gamma rays have also contributed in developing high lipid-producing mutants of *Scenedesmus dimorphic*. On the other hand, UV light mutagenesis generated mutant strains of *Dunaliella bardawil* that are rich in β-carotene, *Phaeodactylum tricornutum* with increased eicosapentaenoic acid production, and *Pavlova lutheri* having higher a EPA and docosahexaenoic acid accumulation, in addition to *Chlorella* strains with better biomass and lipid content (Fu et al. 2016).

Genetic engineering is another approach, in fact a promising one, to develop and optimize algal cell factories such as *Chlamydomonas reinhardtii,* a widely studied microalgae with an established nuclear genetic system (Benedetti et al. 2018). Many of the engineering strategies are summarized in Fig. 3. Novel studies confirmed that expressing *H. pluvialis* β-carotene ketolase in *C. reinhardtii* resulted in ketolutein and adonixanthin production. Alternatively, nuclear transformation of *C. reinhardtii* with phytoene synthase gene from *Chlorella zofingiensis* also increased violaxanthin and lutein yields by 2.0- and 2.2-fold, respectively (Wang et al. 2014).

Moreover, the overexpression of NsbZIP1, a transcription factor transporting the basic leucine zipper, resulted in enhanced higher lipid contents in transformed *Nannochloropsis salina* (Kwon et al. 2017). Conversely in another study, lipid production in *Nannochloropsis gaditana* was doubled by deleting a transcription factor (a homolog of $Zn(II)_2Cys_6$-encoding genes) that operates as negative regulator in lipid biosynthesis (Ajjawi et al. 2017).

Interestingly, a number of algal strains are able of metabolizing agricultural wastes, namely raw scraps of lignocellulosic biomass. However, although these synthesize endogenous Cell Wall Degrading Enzymes (CWDEs), it is not efficient enough for degrading substrates that are hydrolysis-recalcitrant like lignocellulose. Thus, expression of various CWDEs such as cellulases, hemicellulases, ligninases, and polygalacturonases would provide a promising perspective (Benedetti et al. 2018). Recently, engineered *C. reinhardtii* strain, CrAXE03, has managed to synthesize and secrete fungal acetylxylan esterase, which hydrolyzed acetylesters in lignocellulosic biomass by 96%. This has also resulted in an eight-fold increase in the microalgae cell counts with minor growth defects, suggesting that the secretion obliges a limited metabolic burden. Nevertheless, these results encourage the

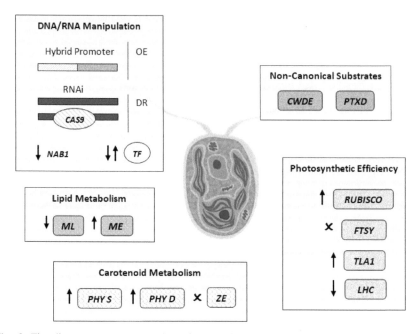

Fig. 3. The diagram presents a number of metabolic engineering strategies, aimed to enhance productivity in *Chlamydomonas reinhardtii* and other microalgal cells. Gene over-expression (OE) using hybrid promoters and gene down-regulation (DR) by RNAi and CRISPR-Cas9 tools are indicated. Other approaches include increased photosynthetic efficiency, the use of non-canonical substrates, as well as optimized carotenoid and lipid metabolism. Besides, the cross refers to loss of function, up- and down-ward pointing arrow refer to up- and down-regulation, in addition to the expression level of the corresponding enzyme (Benedetti et al. 2018).

Abbreviations: *CWDE* cell-wall degrading enzyme, *FTSY* chloroplast signal recognition particle, *LHC* light harvesting complexes, *ME* malate dehydrogenase, *ML* multifunctional lipase, *NAB1* RNA-binding protein, *PHY D* phytoene desaturase, *PHY S* phytoene synthase, *PTXD* phosphite dehydrogenase, *TF* transcription factor, *TLA1* truncated light-harvesting antenna 1, *ZE* zeaxanthin epoxidase.

application of inexpensive renewable feedstock that is an abundant agricultural by-product for microalgae cultivation (Ramos-Martinez et al. 2019).

Molecular biology tools have been also applied to facilitate gene manipulations. Recently, Δ5-elongase and acyl-CoA-dependent Δ6-desaturase genes obtained from *Ostreococcus tauri* were incorporated in *P. tricornutum,* which successfully accumulated omega-3 polyunsaturated fatty acids. *Agrobacterium*-mediated transformation was also shown to increase oil content in engineered *Schizochytrium* and *Parachlorella Kesslerri* (Hamilton et al. 2014). Furthermore, systems biology is an emerging technology for studying cellular metabolism and identifying potential pathways for metabolic engineering targets. To compare transcriptomic patterns from wild-type and engineered strains, genome-scale metabolic model (GSM) can be used. This model is reconstructed for each organism using next-generation sequence data. For example, GSM of other microorganisms has been used to elucidate the mechanism of DHA biosynthesis in *Schizochytrium limacinum,* generating SR21 iCY1170_DHA model. It also reported that the *in silico* addition of malate and citrate led to improved docosahexaenoic acid production by 24.5% and 37.1%, respectively (Ye et al. 2015).

Algal cell factory potentials

Among the prime candidates for future cell factory developments, *Botryococcus braunii, Chlorella vulgaris, Dunaliella salina,* and *Haematococcus pluvialis* have been investigated (Table 3) for various therapeutically effective, functional compounds (Khan et al. 2018).

Carotenoid compounds having antioxidant properties are reported to be produced by *Dunaliella salina* (more than 14% of *β*-carotene in dry biomass) and *Haematococcus pluvialis* (1–8% of astaxanthin as dry biomass). Other species such as *Chlorella* sp. and *Dunaliella* sp. synthesize and secrete polysaccharides with antiviral and antimicrobial actions, at relatively high levels. Microalgae are also producers of anti-inflammatory compounds. *Isochrysis, Nannochloropsis,* and *Pavlova* species produced ω3 PUFA, which has been applied for treating chronic inflammations such as rheumatism. It was also reported that lycopene extracted from

Table 3. Examples of products from microalgal cell factories and their applications (Khan et al. 2018).

Bioactive/value compound	Producer microalgae	Activity/application
β-carotene	*Dunaliella salina*	Role in vision and the immune system
Astaxanthin	*Haematococcus pluvialis*	Antioxidant activity
Sterols mixture	*Euglena gracilis*	Hypo-cholesterolemia, anticancer, and anti-inflammatory
Sitosterol, campesterol and stigmasterol	*Glaucocystophyte* sp.	Pharmaceutical formulation and nutraceuticals
Microcolin-A	*Lyngbya majuscula*	Immunosuppressive agent
Enzyme SOD (superoxide dismutase)	*Anabaena* and *Porphyridium* sp.	Antioxidant
Enzyme carbonic anhydrase	*Isochrysis galbana*	Converts CO_2 into carbonic acid and bicarbonate
Proteins	*Chlorella* sp.	Reduce cholesterol levels by activating cholecystokinin
Vitamins E and C, as well as β-carotene (vitamin A)	*Porphyridium cruentum*	Vitamins
Vitamins A and E, pyridoxine, nicotinic acid thiamine, riboflavin, and biotin	*Dunaliella salina*	Vitamins
Polyunsaturated fatty acids	*Pavlova lutheri*	Suppress cholesterol levels
Eicosapentaenoic acid (EPA) and docosahexaenoic acid (DHA)	*Phaeodactylum tricornutum*	Treatment of inflammatory diseases, heart problems, arthritis, and asthma
Auroxanthin	*Undaria pinnatifida*	Strong radical scavenging action
Fucoxanthin	*Cylindrotheca closterium* and *Phaeodactylum tricornutum*	Anti-cancer, anti-oxidant, anti-obesity, anti-diabetic, and anti-inflammatory agent

Chlorella marina significantly improved rheumatoid arthritis and reduced prostate cancer proliferation in mice (De Morais et al. 2015).

Bioenergy from microalgae

Most micoralgae species are adventitious for biodiesel production because of their high lipid content. For example, *Botryococcus braunii* accumulates up to 80% of oil in its biomass. Carbohydrates such as glycogen, cellulose, and starch are also used for bioethanol production, as they can be easily converted to fermentable sugars. *Chlamydomonas reinhardtii* and *Chlorella vulgaris* are some the carbohydrate rich microalgae, which are considered as potential candidates (Zhu et al. 2016). Recently, it was found that 11.7 g/l final ethanol yield was obtained from *Chlorella vulgaris*. Moreover, the hydrolysis of *Scenedesmus dimorphus* carbohydrate contents (53.7 w/w) with sulfuric acids has resulted in 80% fermentable sugars. Another study reported an ethanol yield of 93% from sugars' fermentation that was derived from *Scenedesmus* sp., indicating this strain's feasibility for biofuel production (Khan et al. 2018).

Microalgae have been utilized as feedstock for producing biohydrogen. To enhance the production efficiency of hydrogen, different partial photosynthetic reactions were manipulated. Using site directed mutagenesis, the RuBisCO activity was downregulated in *C. reinhardtii*, resulting in ten-fold higher H_2 production. It was proposed that Calvin-Benson cycle competed with hydrogenases for reducing equivalents (Benedetti et al. 2018). In addition, the mutant *C. reinhardtii* PFL1 does not produce formate as it lacks PFL enzyme, but does have higher production of light-independent hydrogen. This could be due to decreased expression of [FeFe]-hydrogenase in the mutant type. In another clever approach, a ferredoxin and [FeFe]-hydrogenase fusion gene was expressed in *C. reinhardtii* lacking the native [FeFe]-hydrogenase. This enzyme is coupled to the photosynthetic electron transport (PET) chain and can accept electrons from ferredoxin for H_2 production. Researchers reported that the mutant synthesized 4.5 times more hydrogen than the wild type. On the top of that, further evidence imposes that the engineered protein was more resistant to oxygen (Bayro-Kaiser and Nelson 2017).

Other alternative energy sources such as MFCs can be exploited to recover energy in the form of electricity. As mentioned earlier, these can use microalgae as catalysts to generate bioelectricity, either as electron acceptors in the cathode or donors in the anode. To date, only *Chlamydomonas* and *Chlorella* species have been tested as electricigens. In a novel study, *Chlamydomonas reinhardtii* was investigated in PMFCs by comparing different light intensities. Red LED light was found to produce the highest power density, reaching up to 12.95 mW/m². On the other hand, *Chlorella pyrenoidosa* generated electricity on PMFC's anode without externally added substrates resulting in power density as high as 6030 mW/m² (Cao et al. 2019). Besides, *Chlorella* sp. UMACC 313 is a newly isolated strain that was exploited to form biofilms on the anode, producing 0.124 mW/m² as the highest power output in the PMFC. However, when the microalgal cells were immobilized in alginate gel within the MFC, the maximum power output was enhanced to 0.289 mW/m². This strategy could be the focus of future research for MFCs due to high transparency, biocompatibility, and low toxicity of alginate (Ng et al. 2017).

Fungi

In practice, genetic transformation of filamentous fungi meets with numerous difficulties. For example, currently published protocols contain insufficient information. Also, it is worth mentioning that there exists a vast number of fungal species with complex cell wall structures. Therefore, transformation protocols must be species-specific and optimized for each strain investigated. Establishing completely new methods are also required for efficient transformation (Li et al. 2017).

Fungal CYPs suffer from certain limitations that obstruct their feasible applications. These include lack of stability, low activity, and poor expression levels. Using bioinformatics tools, thermostable fungal CYPs from *Myceliophthora thermophila* and *Thielavia terrestris* were shown to have higher thermal tolerance, and thus can be used to improve the stability of mesophilic CYP enzymes. On the other hand, the oxidizing activity of *CYP5136A3* in *Phanerochaete chrysosporium* was significantly enhanced through site-directed mutagenesis, which also improved expression levels of LovA enzyme in *Aspergillus terreus* (Durairaj et al. 2016).

Despite the progress in bioinformatics tools and genetic engineering to identify BGCs, it is still impractical to identify potentially interesting metabolites. Therefore, future discovery programs must use methods that express foreign pathways and identify the novel compounds using advanced metabolomics (Nielsen and Nielsen 2017). Since such approaches rely on high throughput, more efforts are needed to implement cloning methods to *Penicillium chrysogenum,* which would be high throughput and enable additional studies to utilize the enormous unexploited source for natural products hidden in fungal metagenomes (Guzmán-Chávez et al. 2018a). Similarly, dynamic/kinetic models are required to predict the behavior of *Aspergillus niger* responding to changes during citric acid fermentation. This would aid in the optimization of metabolic engineering design. Besides, high throughput platforms such as strain cultivation, metabolite identification, and fermentation increment, must be developed to test the designed strains in a large-scale (Tong et al. 2019).

Cyanobacteria

Low photosynthetic efficiencies of cyanobacteria present a major challenge for the production of carbon-containing industrial products. Recently, it was demonstrated that over expression of *Synechococcus* sp. PCC 6301 RuBisCO genes in *S. elongatus* PCC 7942 has enhanced CO_2 fixation and boosted isobutanol levels by twofold (Zhou et al. 2016).

Another issue related to these photosynthesizing microbes is the loss of excessive capture of photons, limiting light availability to cells beneath. Light-harvesting antenna size was successfully reduced by engineering a phycocyanin-deficient *Synechocystis* strain, which resulted in higher photosynthetic activity. Moreover, engineering of one of the two photosystems has extended the absorption maxima to 1100 nm (Lau et al. 2015).

Other approaches such as metabolomics, proteomics, and transcriptomes would enhance our knowledge of cyanobacterial biochemistry and molecular biology. For example, a reconstructed metabolic model for *Synechocystis* sp. PCC 6803 revealed that the regulation of photosynthetic activity is far more complex and requires a high degree of co-operability between nine electron flow pathways in order to achieve optimal photoautotrophic metabolism. Transcriptomes studies on engineered *Synechocystis* sp. PCC 6803 have provided insights into understanding the effects of PHA biosynthetic genes over expression on biopolymer production (Al-Haj et al. 2016).

Furthermore, more in-depth investigations of newly reported a glyoxylate shunt and Entner–Doudoroff (ED) pathway will facilitate the evaluation of current and future applied metabolic engineering strategies. Strong, specific promoters are also needed to augment production pathways. Other additional proposed resolutions could include combining natural pathways to create hybrid pathways using *in silico* simulations, so that chemical conversion routes can be efficiently coupled (Zhou et al. 2016).

Microalgae

Although microalgae are valuable sources of bioactive compounds, yet there are still challenges to be addressed. First, open-air systems that are used to grow microalage at a large scale suffer from many problems such as contamination, weather variability, and predators. For this reason, several types of closed photo bioreactors have been developed to operate with minimal contamination risks and under defined optimal conditions, e.g., enhanced light intensity and surface coatings to prevent microalgae adhesion (García et al. 2017). However, scale-up methodologies have to be enhanced in order to achieve minimal carbon dioxide losses, efficient light provisions, and efficient mixing. Continuous culture cultivation, biomass pretreatment, and ensuring maximum yielding fermentation also add new challenges in this field. Nevertheless, more research on constructing inexpensive and ecofriendly photo bioreactors is required to make profitable production of natural compounds derived from microalgae (Khan et al. 2018).

Systems and synthetic biology are areas that also require further attention. New *omics* data are necessary to investigate the pathways involved in the production of microalgal metabolites. More work is still required on the genetic manipulation and transformation of engineered strains to favor a high-level accumulation of target compounds (Fu et al. 2016).

The PSII-independent pathway of hydrogen production is considered as an excellent option for microalgal renewable energy, especially exploiting temperature-sensitive PSII mutants over different temperature cycles. These can be engineered by Cas9-CRISPR technology, which is currently developed. Also, the residual organic matter can be used as feed for a vast number of livestock. Accordingly, integrating these methods into one system would achieve a real sustainable energy conversion (Bayro-Kaiser and Nelson 2017).

Genome editing technologies

Clustered Regularly Interspaced Short Palindromic Repeats (CRISPR)

The manipulation of metabolic pathways can be facilitated by genome editing technologies, which enable both specific gene integration and gene deletion. Among these, CRISPR (clustered regularly interspaced short palindromic repeats) has been successfully developed in various microbial cell factories and was briefly mentioned in this chapter as part of the optimization methodologies.

The implementation of genome editing tools such as zinc-finger nucleases (ZEN), RNA interference (RNAi), and transcription activator-like effecter nucleases (TALEN) in synthetic biology and metabolic engineering is laborious and time-consuming in contrast to the prevailing CRISPR-Cas9 technology (Tian et al. 2017). Recently, a novel approach based on CRISPR-Cas9 was tested in *T. reesei* for the first time. It stimulates gene targeting by introducing specific DNA double-strand breaks, depending only on a *Cas9* nuclease that is CRISPR associated and RNA-guided (see Fig. 4). This accurate targeting was achieved by the protospacer sequence (20 nucleotides) of the guide RNA through simple base pairing. Accordingly, site-specific mutations were generated in target genes via efficient homologous recombination, as well as multiple genes were targeted simultaneously (Bischof et al. 2016). In addition, programming of *Cas9* to identify new targets is much easier compared to TALENs or zinc-finger nucleases, as it only requires a Protospacer Adjacent Motif (PAM) next to the binding site of the guiding protospacer sequence at the target site. This technology efficiently introduced directed mutations into the *yA* gene of *A. nidulans* as well as *albA* and *pyrG* in *Aspergillus aculeatus* (Nødvig et al. 2015). In another study, researchers used CRISPR-Cas9 to disrupt galactaric acid catabolism by deleting several genes in *A. niger*. The resulting strain, which also benefited from the heterologous expression of uronate dehydrogenase, efficiently produced galactaric acid from d-galacturonic acid and from pectin-rich biomass in a consolidated process (Kuivanen et al. 2016).

The CRISPR tool was applied as well in model cyanobacteria including *Anabaena 7120, Synechococcus* 2973, and *Synechocystis* 6803 using Cpf1 from *Francisella novicida* as an alternative nuclease. The *cpf1* genome editing allowed knock-ins, knock-outs, and point mutations in all investigated species. This approach was developed due to the toxicity of Cas9 nuclease in cyanobacteria and the advantages of *cpf1* over *Cas9* nuclease. Some of these include cost effectiveness, cleavage without disrupting PAM and preventing proper editing by homology directed repair, and the requirement of only pre-crRNA to mediate interference instead of both tracrRNA and crRNA. Moreover, in *cpf1* based systems, a single pre-crRNA array containing spacer-repeat sequences can be introduced for multiple targets, whereas in *cas9* systems, a separate tracrRNA-crRNA fusion must be used for every target (Ungerer and Pakrasi 2016). The CRISPR interference (CRISPRi) based repression was also exploited in cyanobacteria due to complexity of genome editing. This tool utilizes catalytically inactive Cas9 (dCas9) and a single guide RNA (sgRNA) to suppress sequence-specific genes exclusive of gene knockout. In *Synechococcus* sp. strain PCC 7002, the pool of α-ketoglutarate was increased after repressing the nitrogen assimilation gene *glnA,* which in turn improved glycolytic flux and lactate production (Mougiakos et al. 2018). Similarly, CRISPRi was employed in *Synechococcus elongatus* PCC 7942 to repress *glgc* (glycogen accumulation), *sdhA* and *sdh*B (succinate conversion to fumarate) genes. This resulted in ameliorated succinate titer by 12.5-fold, without impairing cell growth (Huang et al. 2016).

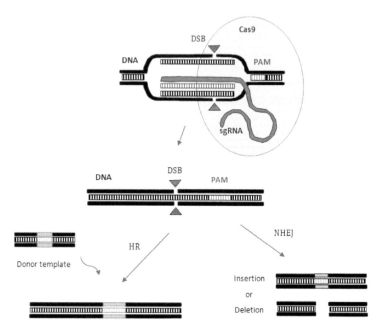

Fig. 4. Schematic representation of CRISPR-Cas9 mediated genome editing. To create a double-strand break (DSB) at 4 nucleotides upstream of the PAM sequence, Cas9 protein forms a complex with sgRNA and binds to the target site of genomic DNA. DSB is repaired by Homologous recombination (HR) or Non-homologous end joining (NHEJ). In HR, specific mutations are integrated into the target genomic site by providing a donor template that has homologous arms with the DSB site. In NHEJ, random insertions or deletions are introduced into the genome (Bischof et al. 2016, Tian et al. 2017).

Multiplex genome editing using CRISPR with Cas9 and Cas12a has been successfully implemented in microorganisms. It allows targeting multiple loci simultaneously by expressing a single CRISPR array, which contains several spacers under the transcriptional regulation of a particular promoter and terminator. Therefore, supplying many targeting expression constructs is not required (Adiego-Pérez et al. 2019). In microalgae, the first CRISPR-Cas9 based genome editing was reported in *C. reinhardtii* where endogenous FKB12 (rapamycin sensitivity) gene was successfully modified regardless of the extreme cytotoxicity of Cas9 towards *C. reinhardtii*. Here, the authors suggested that tactics such as the use of conditional promoters for Cas9 gene expression and transformation with short-lived Cas9 mRNAs might be tested to solve the issue. In a separate study, lipid production in *Nannochloropsis gaditana* was doubled by decreasing expression of the transcription factor Zn_2Cys_6 using CRISPR-Cas9. Although knockout mutants grew poorly, the cytotoxic effects of Cas9 expression were not observed in cultures (Naduthodi et al. 2018). Moreover, multiple gene knock-outs in the diatom *Phaeodactylum tricornutum* were achieved in a single step by delivering six ribonucleoprotein (RNP) complexes, consisting of gRNAs complexed to the recombinant Cas9 protein. This established DNA-free approach identified *PtUMPS* and *PtAPT* genes and validated that their inactivation confers resistance to specific molecules, which could be an alternative to conventional antibiotic-resistance genes (Serif et al. 2018).

Challenges of CRISPR applications

Despite the versatility and feasibility of CRISPR-Cas9 technologies, their application still faces certain challenges. For instance, the expression of *cas9* and sgRNA genes may vary between fungal species due to fixed promoters. If they are poorly expressed, this may lead to inefficient RNA-guided mutagenesis. Nødvig and her colleagues alleviated this problem by substituting the current promoters in their system by strong promoters from the host species (Nødvig et al. 2015). Besides, successful application of CRISPRi in cyanobacteria depends heavily on sgRNA design as selected target gene and binding site could intensely influence the outcome. Thus, future studies should be directed towards developing synthetic components/ routes for strong sgRNA expression and tightly regulatable dCas9 expression (Huang et al. 2016). In multiplexing CRISPR-Cas systems, major challenges include dDNA design and delivery, innovative plasmid assembly methods, and using novel or improved endonucleases. Therefore, dDNA might be stabilized chemically and proper determination of optimal size of homology sequences and other elements is required for the specific host organism. Secondly, smart DNA construction schemes are necessary to develop single or multi-plasmid systems enclosing elements of the CRISPR-Cas system, as well as techniques to incorporate the repeat sequences of CRISPR arrays. Finally, endonucleases are preferred to be smaller, more active, and more specific in addition to having less-stringent PAM recognition selection. Alternatively, the discovery of new *Cas* endonucleases can also extend the PAM compatibility (Adiego-Pérez et al. 2019).

Other challenges that could arise from applying CRISPR-Cas9 with other genetic tools is that CRISPR will not work well for all genetic engineering purposes. For example, the generation of Cas9 depends on translation machinery and thus consumes more resources compared to RNA interference where translation process is not needed. Likewise, CRISPR-Cas9 can trigger off-target mutations that may alter microbial phenotypes. Since these effects are attributed mainly to sgRNAs, they can be designed by harnessing online software such as CHOP-IT and CRISPRdirect (Tian et al. 2017). Last but not the least, advances in CRISPR genome editing of microorganisms can radically accelerate future strain enhancement programs to develop cell factories with exceptional efficiencies. For this reason, dedicated optimization of CRISPR-Cas9 elements is crucial for efficient genome editing and for expanding the quantity of simultaneous editing events. New workflows are also being developed to broaden the knowledge about the mechanism of this system.

Conclusions

In summary, the spectrum of microorganisms exploited as cell factories has expanded from the early prevailing *Escherichia coli* to alternative other bacterial and fungi species. The market and the potential of this recent approach are increasing by taking advantage of a steadily growing range of synthetic biological strategies, as well as newly developed metabolic engineering platforms. Despite the weaknesses of microbial cells as factories compared to other candidates such as mammalian cell lines, they are still favored due to their versatility, eco-friendliness, and cost-effective cultivation. Hence, challenges in the field of cell factory development, especially for newly discovered and exploited species, must be addressed in numerous ways. Fortunately, genome-editing tools such as CRISPR hold great promises for facilitating coherent and efficient design of biological systems, thus accelerating cell factory engineering.

Future perspective

Considering the great potentials of diverse microbial cell factories in many biotechnological applications, it would be of industrial, scientific, and technological interests to translate these minute systems' potential into possible large-scale production. Although a wide-ranging spectrum of natural products biosynthesized by microbes has potential industrial applications, the underlying challenges in commercialization are to accomplish production yields that meet demands of reasonable scalable production.

Acknowledgments

Author wishes to thank family and professors from Qatar University for their encouragements to write this chapter contributing to the book.

References

Adiego-Pérez, B., Randazzo, P., Daran, J. M., Verwaal, R., Roubos, J. A., Daran-Lapujade, P. et al. 2019. Multiplex genome editing of microorganisms using CRISPR-Cas. FEMS Microbiology Letters 366(8): fnz086. doi:10.1093/femsle/fnz086.

Al-Haj, L., Lui, Y. T., Abed, R. M., Gomaa, M. A. and Purton, S. 2016. Cyanobacteria as chassis for industrial biotechnology: progress and prospects. Life (Basel, Switzerland) 6(4): 42. doi:10.3390/life6040042.

Ajjawi, I., Verruto, J., Aqui, M., Soriaga, L. B., Coppersmith, J., Kwok, K. et al. 2017. Lipid production in *Nannochloropsis gaditana* is doubled by decreasing expression of a single transcriptional regulator. Nature Biotechnology 35(7): 647–652. doi:10.1038/nbt.3865.

Bayro-Kaiser, V. and Nelson, N. 2017. Microalgal hydrogen production: Prospects of an essential technology for a clean and sustainable energy economy. Photosynthesis Research 133(1-3): 49–62. doi:10.1007/s11120-017-0350-6.

Benedetti, M., Vecchi, V., Barera, S. and Dall'Osto, L. 2018. Biomass from microalgae: the potential of domestication towards sustainable biofactories. Microbial cell Factories 17(1): 173. doi:10.1186/s12934-018-1019-3.

Bischof, R. H., Ramoni, J. and Seiboth, B. 2016. Cellulases and beyond: the first 70 years of the enzyme producer *Trichoderma reesei*. Microbial Cell Factories 15(1): 106. doi:10.1186/s12934-016-0507-6.

Bond, C., Tang, Y. and Li, L. 2016. *Saccharomyces cerevisiae* as a tool for mining, studying and engineering fungal polyketide synthases. Fungal Genetics and Biology: FG & B 89: 52–61. doi:10.1016/j.fgb.2016.01.005.

Calero, P. and Nikel, P. I. 2018. Chasing bacterial chassis for metabolic engineering: a perspective review from classical to non-traditional microorganisms. Microbial Biotechnology 12(1): 98–124. doi:10.1111/1751-7915.13292.

Cao, Y., Mu, H., Liu, W., Zhang, R., Guo, J., Xian, M. et al. 2019. Electricigens in the anode of microbial fuel cells: pure cultures versus mixed communities. Microbial Cell Factories 18(1): 39. doi:10.1186/s12934-019-1087-z.

Davy, A. M., Kildegaard, H. F. and Andersen, M. R. 2017. Cell factory engineering. Cell Systems 4(3): 262–275. doi:10.1016/j.cels.2017.02.010.

De Morais, M. G., Vaz, B., de Morais, E. G. and Costa, J. A. 2015. Biologically active metabolites synthesized by microalgae. BioMed Research International 2015: 835761. doi:10.1155/2015/835761.

Druzhinina, I. S. and Kubicek, C. P. 2017. Genetic engineering of *Trichoderma reesei* cellulases and their production. Microbial Biotechnology 10(6): 1485–1499. doi:10.1111/1751-7915.12726.

Durairaj, P., Hur, J. S. and Yun, H. 2016. Versatile biocatalysis of fungal cytochrome P450 monooxygenases. Microbial Cell Factories 15(1): 125. doi:10.1186/s12934-016-0523-6.

Eungrasamee, K., Miao, R., Incharoensakdi, A., Lindblad, P. and Jantaro, S. 2019. Improved lipid production via fatty acid biosynthesis and free fatty acid recycling in engineered *Synechocystis* sp. PCC 6803. Biotechnology for Biofuels 12: 8. doi:10.1186/s13068-018-1349-8.

Fu, W., Chaiboonchoe, A., Khraiwesh, B., Nelson, D. R., Al-Khairy, D., Mystikou, A. et al. 2016. Algal cell factories: Approaches, applications, and potentials. Marine Drugs 14(12): 225. doi:10.3390/md14120225.

García, J. L., de Vicente, M. and Galán, B. 2017. Microalgae, old sustainable food and fashion nutraceuticals. Microbial Biotechnology 10(5): 1017–1024. doi:10.1111/1751-7915.12800.

Grijseels, S., Pohl, C., Nielsen, J. C., Wasil, Z., Nygård, Y., Nielsen, J. et al. 2018. Identification of the decumbenone biosynthetic gene cluster in *Penicillium decumbens* and the importance for production of calbistrin. Fungal Biology and Biotechnology 5: 18. doi:10.1186/s40694-018-0063-4.

Guzmán-Chávez, F., Salo, O., Nygård, Y., Lankhorst, P. P., Bovenberg, R. and Driessen, A. 2017. Mechanism and regulation of sorbicillin biosynthesis by *Penicillium chrysogenum*. Microbial Biotechnology 10(4): 958–968. doi:10.1111/1751-7915.12736.

Guzmán-Chávez, F., Zwahlen, R. D., Bovenberg, R. and Driessen, A. 2018a. Engineering of the filamentous fungus *Penicillium chrysogenum* as cell factory for natural products. Frontiers in Microbiology 9: 2768. doi:10.3389/fmicb.2018.02768.

Guzman-Chavez, F., Salo, O., Samol, M., Ries, M., Kuipers, J., Bovenberg, R. et al. 2018b. Deregulation of secondary metabolism in a histone deacetylase mutant of *Penicillium chrysogenum*. MicrobiologyOpen 7(5): e00598. doi:10.1002/mbo3.598.

Häkkinen, M., Valkonen, M. J., Westerholm-Parvinen, A., Aro, N., Arvas, M., Vitikainen, M. et al. 2014. Screening of candidate regulators for cellulase and hemicellulase production in *Trichoderma reesei* and identification of a factor essential for cellulase production. Biotechnology for Biofuels 7(1): 14. doi:10.1186/1754-6834-7-14.

Hamilton, M. L., Haslam, R. P., Napier, J. A. and Sayanova, O. 2014. Metabolic engineering of *Phaeodactylum tricornutum* for the enhanced accumulation of omega-3 long chain polyunsaturated fatty acids. Metabolic Engineering 22(100): 3–9. doi:10.1016/j.ymben.2013.12.003.

Hidalgo, P. I., Ullán, R. V., Albillos, S. M., Montero, O., Fernández-Bodega, M. Á., García-Estrada, C. et al. 2014. Molecular characterization of the PR-toxin gene cluster in *Penicillium roqueforti* and *Penicillium chrysogenum*: Cross talk of secondary metabolite pathways. Fungal Genetics and Biology 62: 11–24. doi:10.1016/j.fgb.2013.10.009.

Huang, C. H., Shen, C. R., Li, H., Sung, L. Y., Wu, M. Y. and Hu, Y. C. 2016. CRISPR interference (CRISPRi) for gene regulation and succinate production in cyanobacterium *S. elongatus* PCC 7942. Microbial Cell Factories 15(1): 196. doi:10.1186/s12934-016-0595-3.

Huccetogullari, D., Luo, Z. W. and Lee, S. Y. 2019. Metabolic engineering of microorganisms for production of aromatic compounds. Microbial Cell Factories 18(1): 41. doi:10.1186/s12934-019-1090-4.

Jindou, S., Ito, Y., Mito, N., Uematsu, K., Hosoda, A. and Tamura, H. 2014. Engineered platform for bioethylene production by a cyanobacterium expressing a chimeric complex of plant enzymes. ACS Synthetic Biology 3: 487–496.10.1021/sb400197f.

Kavšček, M., Stražar, M., Curk, T., Natter, K. and Petrovič, U. 2015. Yeast as a cell factory: current state and perspectives. Microbial Cell Factories 14: 94. doi:10.1186/s12934-015-0281-x.

Khan, M. I., Shin, J. H. and Kim, J. D. 2018. The promising future of microalgae: current status, challenges, and optimization of a sustainable and renewable industry for biofuels, feed, and other products. Microbial Cell Factories 17(1): 36. doi:10.1186/s12934-018-0879-x.

Knoot, C. J., Ungerer, J., Wangikar, P. P. and Pakrasi, H. B. 2018. Cyanobacteria: Promising biocatalysts for sustainable chemical production. The Journal of Biological Chemistry 293(14): 5044–5052. doi:10.1074/jbc.R117.815886.

Knoot, C. J. and Pakrasi, H. B. 2019. Diverse hydrocarbon biosynthetic enzymes can substitute for olefin synthase in the cyanobacterium *Synechococcus* sp. PCC 7002. Scientific Reports 9(1): 1360. doi:10.1038/s41598-018-38124-y.

Krause, M., Neubauer, A. and Neubauer, P. 2016. The fed-batch principle for the molecular biology lab: controlled nutrient diets in ready-made media improve production of recombinant proteins in *Escherichia coli*. Microbial Cell Factories 15(1): 110. doi:10.1186/s12934-016-0513-8.

Kuivanen, J., Wang, Y. J. and Richard, P. 2016. Engineering *Aspergillus niger* for galactaric acid production: elimination of galactaric acid catabolism by using RNA sequencing and CRISPR/Cas9. Microbial Cell Factories 15(1): 210. doi:10.1186/s12934-016-0613-5.

Kumaraswamy, K. G., Krishnan, A., Ananyev, G., Zhang, S., Bryant, D. and Dismukes, G. 2019. Crossing the thauer limit: Rewiring cyanobacterial metabolism to maximize fermentative H_2 production. Energy & Environmental Science 12(3). doi:10.1039/c8ee03606c.

Kwon, S., Kang, N. K., Koh, H. G., Shin, S., Lee, B., Jeong, B. et al. 2017. Enhancement of biomass and lipid productivity by overexpression of a bZIP transcription factor in *Nannochloropsis salina*. Biotechnology and Bioengineering 115(2): 331–340. doi:10.1002/bit.26465.

Lau, N. S., Matsui, M. and Abdullah, A. A. 2015. Cyanobacteria: Photoautotrophic microbial factories for the sustainable synthesis of industrial products. BioMed. Research International 2015: 754934. doi:10.1155/2015/754934.

Li, D., Tang, Y., Lin, J. and Cai, W. 2017. Methods for genetic transformation of filamentous fungi. Microbial Cell Factories 16(1): 168. doi:10.1186/s12934-017-0785-7.

Liu, X., Ding, W. and Jiang, H. 2017. Engineering microbial cell factories for the production of plant natural products: from design principles to industrial-scale production. Microbial Cell Factories 16(1): 125. doi:10.1186/s12934-017-0732-7.

Lv, X., Zheng, F., Li, C., Zhang, W., Chen, G. and Liu, W. 2015. Characterization of a copper responsive promoter and its mediated overexpression of the xylanase regulator 1 results in an induction-independent production of cellulases in *Trichoderma reesei*. Biotechnology for Biofuels 8: 67. doi:10.1186/s13068-015-0249-4.

Madsen, M. A., Semerdzhiev, S., Amtmann, A. and Tonon, T. 2018. Engineering mannitol biosynthesis in *Escherichia coli* and *Synechococcus* sp. PCC 7002 using a green algal fusion protein. ACS Synthetic Biology 7(12): 2833–2840. doi:10.1021/acssynbio.8b00238.

Maheshwari, D. K., Dubey, R. C. and Saravanamuthu, R. 2010. Industrial Exploitation of Microorganisms. New Delhi: I.K. International Publishing House.

Mougiakos, I., Bosma, E. F., Ganguly, J., Oost, J. V. and Kranenburg, R. V. 2018. Hijacking CRISPR-Cas for high-throughput bacterial metabolic engineering: Advances and prospects. Current Opinion in Biotechnology 50: 146–157. doi:10.1016/j.copbio.2018.01.002.

Naduthodi, M. I., Barbosa, M. J. and Van Der Oost, J. 2018. Progress of CRISPR-Cas based genome editing in photosynthetic microbes. Biotechnology Journal 13(9): e1700591. doi:10.1002/biot.201700591.

Negrete, A. and Shiloach, J. 2017. Improving *E. coli* growth performance by manipulating small RNA expression. Microbial Cell Factories 16(1): 198. doi:10.1186/s12934-017-0810-x.

Ng, F. L., Phang, S. M., Periasamy, V., Yunus, K. and Fisher, A. C. 2017. Enhancement of power output by using alginate immobilized algae in biophotovoltaic devices. Scientific Reports 7(1): 16237. doi:10.1038/s41598-017-16530-y.

Nielsen, J. C. and Nielsen, J. 2017. Development of fungal cell factories for the production of secondary metabolites: Linking genomics and metabolism. Synthetic and Systems Biotechnology 2(1): 5–12. doi:10.1016/j.synbio.2017.02.002.

Nielsen, J. 2019. Cell factory engineering for improved production of natural products. Natural Product Reports. doi:10.1039/c9np00005d.

Nødvig, C. S., Nielsen, J. B., Kogle, M. E. and Mortensen, U. H. 2015. A CRISPR-Cas9 system for genetic engineering of filamentous fungi. PloS One 10(7): e0133085. doi:10.1371/journal.pone.0133085.

Pohl, C., Mózsik, L., Driessen, A. J. M., Bovenberg, R. A. L. and Nygård, Y. I. 2018. Genome editing in *Penicillium chrysogenum* using Cas9 ribonucleoprotein particles. Methods in Molecular Biology 213–232. doi:10.1007/978-1-4939-7795-6_12.

Ramos-Martinez, E. M., Fimognari, L., Rasmussen, M. K. and Sakuragi, Y. 2019. Secretion of acetylxylan esterase from *Chlamydomonas reinhardtii* enables utilization of lignocellulosic biomass as a carbon source. Frontiers in Bioengineering and Biotechnology 7: 35. doi:10.3389/fbioe.2019.00035.

Roell, G. W., Zha, J., Carr, R. R., Koffas, M. A., Fong, S. S. and Tang, Y. J. 2019. Engineering microbial consortia by division of labor. Microbial Cell Factories 18(1): 35. doi:10.1186/s12934-019-1083-3.

Sanchez-Garcia, L., Martín, L., Mangues, R., Ferrer-Miralles, N., Vázquez, E. and Villaverde, A. 2016. Recombinant pharmaceuticals from microbial cells: a 2015 update. Microbial Cell Factories 15: 33. doi:10.1186/s12934-016-0437-3.

Salo, O. V., Ries, M., Medema, M. H., Lankhorst, P. P., Vreeken, R. J., Bovenberg, R. A. et al. 2015. Genomic mutational analysis of the impact of the classical strain improvement program on β-lactam producing *Penicillium chrysogenum*. BMC Genomics 16: 937. doi:10.1186/s12864-015-2154-4.

Serif, M., Dubois, G., Finoux, A. L., Teste, M. A., Jallet, D. and Daboussi, F. 2018. One-step generation of multiple gene knock-outs in the diatom *Phaeodactylum tricornutum* by DNA-free genome editing. Nature Communications 9(1): 3924. doi:10.1038/s41467-018-06378-9.

Sharma, D. and Saharan, B. S. 2018. Microbial Cell Factories. Boca Raton, FL: CRC Press, Taylor & Francis Group.

Sun, X., Wu, H., Zhao, G., Li, Z., Wu, X., Liu, H. et al. 2018. Morphological regulation of *Aspergillus niger* to improve citric acid production by chsC gene silencing. Bioprocess and Biosystems Engineering 41(7): 1029–1038. doi:10.1007/s00449-018-1932-1.

Tian, P., Wang, J., Shen, X., Rey, J. F., Yuan, Q. and Yan, Y. 2017. Fundamental CRISPR-Cas9 tools and current applications in microbial systems. Synthetic and Systems Biotechnology 2(3): 219–225. doi:10.1016/j.synbio.2017.08.006.

Tong, Z., Zheng, X., Tong, Y., Shi, Y. C. and Sun, J. 2019. Systems metabolic engineering for citric acid production by *Aspergillus niger* in the post-genomic era. Microbial Cell Factories 18(1): 28. doi:10.1186/s12934-019-1064-6.

Ungerer, J. and Pakrasi, H. B. 2016. Cpf1 is a versatile tool for CRISPR genome editing across diverse species of cyanobacteria. Scientific Reports 6: 39681. doi:10.1038/srep39681.

Viggiano, A., Salo, O., Ali, H., Szymanski, W., Lankhorst, P. P., Nygård, Y. et al. 2018. Pathway for the biosynthesis of the pigment chrysogine by *Penicillium chrysogenum*. Applied and Environmental Microbiology 84(4): e02246–17. doi:10.1128/AEM.02246-17.

Wang, C., Kim, J. H. and Kim, S. W. 2014. Synthetic biology and metabolic engineering for marine carotenoids: new opportunities and future prospects. Marine Drugs 12(9): 4810–4832. doi:10.3390/md12094810.

Wang, L., Cao, Z., Hou, L., Yin, L., Wang, D., Gao, Q. et al. 2016. The opposite roles of agdA and glaA on citric acid production in *Aspergillus niger*. Applied Microbiology and Biotechnology 100(13): 5791–5803. doi:10.1007/s00253-016-7324-z.

Xue, Y., Zhang, Y., Cheng, D., Daddy, S. and He, Q. 2014. Genetically engineering *Synechocystis* sp. pasteur culture collection 6803 for the sustainable production of the plant secondary metabolite *P*-coumaric acid. Proceedings of the National Academy of Sciences of the United States of America 111(26): 9449–9454. doi:10.1073/pnas.1323725111.

Xue, Y. and He, Q. 2015. Cyanobacteria as cell factories to produce plant secondary metabolites. Frontiers in Bioengineering and Biotechnology 3: 57. doi:10.3389/fbioe.2015.00057.

Ye, C., Qiao, W., Yu, X., Ji, X., Huang, H., Collier, J. L. et al. 2015. Reconstruction and analysis of the genome-scale metabolic model of *Schizochytrium limacinum* SR21 for docosahexaenoic acid production. BMC Genomics 16: 799. doi:10.1186/s12864-015-2042-y.

Yin, X., Shin, H. D., Li, J., Du, G., Liu, L. and Chen, J. 2017. Comparative genomics and transcriptome analysis of *Aspergillus niger* and metabolic engineering for citrate production. Scientific Reports 7: 41040. doi:10.1038/srep41040.

Yuan, S. F. and Alper, H. S. 2019. Metabolic engineering of microbial cell factories for production of nutraceuticals. Microbial Cell Factories 18(1): 46. doi:10.1186/s12934-019-1096-y.

Zhou, J., Zhu, T., Cai, Z. and Li, Y. 2016. From cyanochemicals to cyanofactories: a review and perspective. Microbial Cell Factories 15: 2. doi:10.1186/s12934-015-0405-3.

Zhu, L. D., Li, Z. H. and Hiltunen, E. 2016. Strategies for lipid production improvement in microalgae as a biodiesel feedstock. BioMed. Research International 2016: 8792548. doi:10.1155/2016/8792548.

CHAPTER 5

Metabolic Engineering

Duygu Nur Arabaci,[1] *Onat Cinlar,*[2] *Aiyoub Shahi*[3] *and*
Sevcan Aydin[4,*]

Introduction

Genetic manipulation and alteration is as old as life; nature solves the problem of surviving and enduring ages and changing conditions through evolution. Most of the alterations happen in the regulation of pathways and structure of enzymes, which provide functionality to life, allowing it to collect and utilize what is around and build those materials into fully functioning organisms. In modern times, humans have been able to intentionally manipulate such mechanisms in order to solve ancient problems as well as the problems modern times have brought.

Metabolic engineering is the discipline that involves the design, construction and optimization of metabolic pathways of microbial systems for metabolite production for various purposes. The core concept of metabolic engineering is manipulation of organisms that can produce desirable outcomes, and this is mainly achieved using molecular biology techniques. Importance of this discipline comes from its flexibility and therefore its applicability to various important issues of modern societies.

Waste treatment

One of the potential applications of metabolic engineering is waste treatment, especially engineering new processes that not only manage waste, but also recycle valuable materials from the process, and these are being studied to fight the pollution of conventional energy resources. The materials that are obtained from waste

[1] Department of Genetics and Bioengineering, Nişantaşı University, Maslak, 34469, Istanbul, Turkey.
[2] Department of Biomedical Engineering, Bahcesehir University, Besiktas, 34353 Istanbul Turkey.
[3] Institute of Environment, University of Tabriz, Tabriz, Iran.
[4] Department of Genetics and Bioengineering, Nişantaşı University, Maslak, 34469, Istanbul, Turkey.
* Corresponding author: sevcan_aydn@hotmail.com

treatment are mostly sugar or glucose based feedstock, which can be transformed into polymers and bulk or fine chemicals (Otero and Nielsen 2010) and many more.

Among the valuable chemicals that can be extracted during waste treatment, arguably one of the most important ones is biogas. Biogas is the result of anaerobic digestion of organic materials, consisting mainly of methane and carbon dioxide, with small amounts of other gases. Biogas, and its modified components, are promising alternatives in the effort to find a sustainable energy source since they can be obtained from different kinds of organic waste, which can include agricultural waste, manure, sewage or municipal wastes. There are various kinds of biogas that can be obtained via different treatment technologies, such as biohydrogen, biomethane and biohythane, which is the combination of biohydrogen and biomethane. Biohythane is a newer product, and because it is formed by methane and hydrogen, it shows extensive potential as vehicle fuel, and has various advantages over conventional fuels, such as lower greenhouse gas emissions and higher fuel efficiency (Pasupuleti and Venkata Mohan 2015). Furthermore, the effluent wastewaters of refineries that treat waste with a secondary goal to obtain biohythane contain valuable materials. Liu et al. (2018) suggest that providing hythane in a sustainable and green way can be made possible with a process of two-stage anaerobic fermentation. However, the stability of the bioreactors for the production of hydrogen and the overall energy efficiency of such systems, as well as the profitability, remain as challenges in the way of replacing fossil fuels with biohythane.

Biohydrogen, as one of the most promising biofuels, has the most energy content per unit weight element, because hydrogen atoms are not bound by carbon atoms, and biohydrogen obtained from waste treatment processes can be processed into various products, mainly into hydrogen and volatile fatty acids (Sakar et al. 2016). In processes that produce biogas from waste, different microbial groups in mixed consortia can be used as biocatalysts (Nikhil et al. 2014). Through such methods, the processes can become more cost-effective and in the future could even push this technique over its chemical and electrolytic counterparts.

Novel methods for wastewater treatment are required to provide solutions for various issues with conventional wastewater treatment. One such problem is called eutrophication phenomenon. This phenomenon is caused by enrichment of aquatic environments with various nutrients but mainly through phosphates and nitrates. According to Cai (2013), these increased nutrients give rise to algal bloomings, over-spreading of aquatic plants and other effects that may damage the aquatic ecosystem. This phenomenon is traditionally solved with biological treatments, followed with various nitrification and denitrification processes (Renuka et al. 2013). Conventional methods like these have several challenges and associated disadvantages to consider.

To overcome these drawbacks, studies are performed with a focus on microalgae in the last few decades. As microalgae consume heavy amounts of nitrogen and phosphorus in the process of their growth, they can easily be implemented in wastewater treatment processes as a counter for eutrophication phenomenon (Renuka et al. 2013, Muñoz and Guieysse 2006). Through microalgae systems, phosphorus and nitrogen can be recycled to be utilized as fertilizers or feedstock for bioenergy applications.

While many applications for wastewater treatment appear in different areas, some considerations must be taken with such methods. Higher process costs, longer process times and increased complexity of procedures could cause a challenge important enough for some projects with limited resources than other methods that they may have to opt for more suitable alternatives.

Utilization of microalgae brings several advantages over conventional wastewater treatment procedures. Unlike in conventional methods, phosphorus and nitrogen can be recycled through utilization of these microalgae into fertilizers, food or bioenergy applications as well as in more areas (Aslan and Kapdan 2006, Rawat et al. 2011, Renuka et al. 2013).

Microbial cell factories

Another important application of metabolic engineering is microbial cell factories. In this approach, cells are differentiated in order to yield products that they would not originally yield. Utilizing this aspect of metabolic engineering enables production of fine and bulk chemical and biological substances. In order to produce such materials, microbial cell factories are manipulated to alter their metabolic pathways, both through genetic manipulation and environmental optimization (Jakočiunas 2016).

L-amino acids are one of the many examples of biochemicals that can be produced with microbial cell factories. While this method has been around for over half a century, new advancements are continuously arising in this area, especially optimizations of microbial cell factories genomes for higher efficiency functions. According to Kalinowski et al. (2003), one of the microorganisms that draw considerable attention in this field for amino acid production is the *Corynebacterium glutamicum*. It is reported to utilize mixed carbon sources simultaneously, namely acetate and glucose, and shows nondiauxic growth pattern, which is an important property that makes it appealing to industrial applications (Wendisch 2007, Bott and Eggeling 2005).

Metabolic engineering methods can be applied to many diverse fields, in which different kinds of microorganisms can be utilized for different purposes. According to Archer (2000), fungi can be used for production of food ingredients. While fungi have been used as cell factories for beverages and various fermented foods such as ales and cheese for a very long time, production of food ingredients is still continuously improving, especially through intentional manipulation of fungi genes. Many studies in this area utilize *Saccharomyces cerevisiae*. This yeast is used in various areas of production, and other filamentous fungi are also being engineered for similar purposes. However, research on utilizing filamentous fungi as microbial cell factories is progressing slowly, mainly because of the disappointing yields of their utilization as hosts for recombinant DNA for the purpose of heterologous enzyme production. In addition to that, their characteristics are not well known, with only a handful of strains being studied commonly (Archer 2000). As advancements in the field of molecular biology make it possible to reveal more and more on the inner machinery of these microorganisms, new and exciting possibilities for production processes open up, promising novel methods to be more efficient than current methods.

A common method to create microbial cell factories using hosts that don't naturally support the intended pathway is artificially installing the pathways required for the final metabolite production. This process may be used to obtain the products that occur naturally in organisms that are hard to culture and engineer, or for products that are derived from higher organisms.

There are many factors to consider when utilizing microbial cell factories. One such factor that takes effect when designing microbial systems as cell factories is the end product. Different products' production efficiencies may differentiate vastly depending on which organism was used as the microbial cell factory (Fisher et al. 2014), as these intended products may not be native metabolites to the host organism. Aside from nativity, various products may require different kinds of metabolic capabilities for the host organism to produce the interested line.

Another important factor that can have an immense effect on the yield of microbial cell factories is the efficiency of the cell factories that are used to produce the end product. Because host selections can vary vastly for various end products, there are additional parameters that should be considered. First of these is the complexity of the required alterations and engineering on the chassis as well as its environment and components of the process. Matching the most efficient chassis for the intended end product is the key for drawing a path for the most efficient process.

Another factor to be considered is the characteristics of the metabolite to be produced, and the potential effects on the host organism. For example, toxicity of the product and its accumulation in the production environment can affect the host, as well as the other components and steps of the process of production. With the huge pool of available cell factories to modify and utilize in order to produce various kinds of chemical and biological products, it is very important to match the most suitable cell factory to produce the specific intended product.

In synthetic biology, chassis organism refers to the host organism that contains the required chemicals and produces the intended end product. In the case of producing non-native metabolites in the chassis organism, there may be a shortage of other native enzymes. This lack of enzymes may lead to errors during pathway cascades and may even have an impact on the pathway components to such extent that it may not function at all; however, they don't always pose a challenge as in many cases the addition of applicable enzymes to the chassis is enough to aid the reaction and resolve the issue.

Microbial cell factories can be implemented for various industrial processes, providing important metabolites. With artificial *de novo* and heterologous biosynthetic pathways gaining popularity, suitable selection of products and processes is becoming increasingly important for profitability. Each and every metabolite to be produced requires thorough research, with different sets of host or chassis, pathways, enzymes and other variables that needs to considered; however, the effectiveness and the potential for increasing efficiency and variability of utilizing microorganisms for production encourages researchers to overcome these weaknesses.

Bioremediation

Developing effective ways for managing biological contaminants is a topic that attracts attention from multiple disciplines. Many studies in this field focus on different aspects and potentials, from efficiency on microalgae production to bioremediation (Loera-Quezada et al. 2016, Kumar et al. 2011). Kamaludeen et al. (2003) reported that the increasing levels of xenobiotic compounds in the environment coupled with scarcity of clean water caused by hazardous waste has led to restrained crop yield around the world. Microbial systematics including different microbial consortia such as fungi, yeast and bacteria offer promising solutions to the rising environmental issues in the modern times (Strong and Burgess 2008). For such purposes, microorganisms are implemented to clean the contaminated soil and water, and there are various researches on such applications. Many researchers study microorganisms that are native to the contaminated areas for ease of control and increased efficiency, as well as an already established ecosystem with decreased unknown factors. Through such studies, various methods and processes are proposed to solve such environmental problems. After introducing microorganisms to the area, their growth is encouraged with the addition of necessary nutrients, thermal electron acceptors and establishing required moisture and temperature. After they reach a significant mass, they are intended to consume the contaminants as nutrients (Hess et al. 1997). In another aspect, the processes behind the utilization of microorganisms as bioremediators varies depending on the contaminant. For example, these procedures may target specific contaminants, or the procedure may implement a more wholesome approach, which would require different approaches.

Bioremediation offers significant advantages in resolving environmental problems, but many challenges remain. Some heavy metals such as lead might require extra processes because microorganisms cannot integrate these metals readily (Vidali 2001), while some of them such as mercury cannot be processed at all, because accumulation of mercury in these microorganisms can lead to integration of mercury to food chain, causing more problems.

Another challenge is that bioremediation site conditions may not be suitable for utilization of native strains, and in those cases genetically engineered bioremediation systems can be utilized to allow the site to support the intended microorganisms. Such bioremediation systems are purposefully introduced into the site, and they accelerate the degradation process by propagating the population of microorganism. This process can include different substances such as oxygen, nitrogen, phosphorous, electron acceptors, and nutrients (Höhener et al. 1998). For this kind of bioremediation processes, main goals are mineralization of contaminants, increased rates for the processes, availability to be utilized in various regions, the ability for the bioremediation process to resume functionality with continuous influent waste and total transformation into harmless products. Compared to other bioremediation techniques, engineered *in situ* methods don't require high energy input and conserve soil structure during the process of detoxification. Even though it has many important advantages over other methods, it takes a long time to take effect. So engineered *in situ* bioremediation is not a silver bullet for environmental issues.

Climate change

In Anthropocene, or the age when humans have a massive impact on the planet, the results are accumulating. In all levels of production, from harvest to disposal, more and more chemicals are being released into the waters, the oceans and the atmosphere. With the population also on the rise, the sustenance required to support humans is ever increasing, which creates a downward spiral of more demand, more production, and a requirement for more resources and means of disposal. According to OECD, the global water demand is estimated to increase by 55% from 2000 to 2050, of which 70% is used on agriculture, and the food production is estimated to expand 60% by 2035 to support the increasing population. However, Earth contains only limited resources, and according to a 2015 study, they are decreasing faster than they can be replenished, with 21 out of 37 major underground fresh water reservoirs receding (Richey et al. 2015). While the planet contains 80% water, 97.5% of it is unfit for human consumption, leaving only 2.5% that can be used for drinking, agriculture and various industrial applications. Moreover, the rising concentrations of CO_2 in the atmosphere may increase the levels of tropospheric ozone, which can cause damage to living tissues (Amann 2008).

One of the most important environmental issues of the 21st century has been climate change. Climate is defined as the average weather patterns of a region over long periods of time, and globally, climate is being hugely affected by anthropogenic activity, especially through the emission of greenhouse gases such as CO_2, CH_4, N_2O and various fluorinated gases. These gases allow sunlight to pass through, but absorb the heat reflected back from the Earth's surface, therefore trapping heat within the atmosphere, this phenomena is called the greenhouse effect. The temperature increase caused by increased accumulation of these gases leads to positive feedback loops to amplify these effects; water vapor levels increase due to the warmer temperatures, which absorbs more infrared light, glaciers melt, therefore the icy reflective surface of the oceans are replaced with darker waters that absorb more light. Climate systems are complex and interdependent across the globe, and due to these positive feedback loops, and even small changes, such as the greenhouse gas emissions by human activity that only account for 5% of global warming, lead to huge impacts that may lead to a point of no return.

Microbial communities are known to play a crucial role in the cycles that regulate greenhouse gases in the atmosphere, such as the carbon cycle (Lal 2008). With the changes occurring within the concentration of these gases within the atmosphere and the following increase in temperature, many studies have found that the microbial communities are being disturbed, their compositions and diversity and thus their metabolic activities that relate to greenhouse gas emissions are being altered (Hungate et al. 1997, Smith et al. 2008, Singh et al. 2010, Zhou et al. 2011). It is widely accepted that microbial processes have a huge impact on climate change and are also hugely affected by it. Therefore, the manipulation of microbial ecosystems remains a promising tool for intervention towards climate change. The climate change mostly affects the microbial communities indirectly, mainly through altered plant and soil properties. Most microorganisms carry out biological processes and grow faster in higher temperatures; therefore, the rising temperatures globally may lead to faster rates of methanogenesis, respiration and fermentation, which are the main processes that release greenhouse gases into the atmosphere in the overall scale (Smith et al. 2008). Also, since different microbial species have different optimal temperatures, the faster rates of growth of some microbial communities may lead to them dominating the other species, altering the composition of the community and loss of diversity, potentially the elimination of functions, for example, the population of bacteria that carry out nitrification may be lost all together (Delgado-Baquerizo et al. 2017).

Most of the atmospheric CO_2 fluxes mostly depend on respiration and photosynthesis. Across the oceans, phytoplankton is responsible for the majority of photosynthesis, and through autotrophic and heterotrophic respiration, the carbon that was taken up during photosynthesis is returned to the dissolved inorganic carbon pool. On land surfaces, while plants take up the majority of CO_2 from the atmosphere, microorganisms directly play a part in the carbon cycle through decomposition of decaying matter and heterotrophic respiration, or indirectly by modification of nutrient availability in soil and acting as pathogens, symbionts or even through root herbivores, affecting plant metabolic processes (Bardgett and van der Putten 2014).

While natural emissions of methane are mostly due to methanogenesis carried out by anaerobic archaea underwater in oceans or in marshes, as well as in the gastrointestinal tracts of animals, the emissions due to anthropogenic activity outweigh the natural processes, majority of which are dependent upon microbial communities. Methanotrophic bacteria consume CH_4 in the atmosphere, or degrade it before it can reach there, and an estimated 30 million tons of methane is taken out of the atmosphere per year by high affinity methanotrophs, majority of which are made up of Alphaproteobacteria, or type II methanotrophs (IPCC 2007).

Microorganisms are also the main mediators for N_2O cycles, especially in soil, through processes of nitrification and denitrification. Majority of N_2O released from the soil via nitrification processes is due to the processes being carried out by Betaproteobacteria, which are ammonia-oxidizing bacteria, in the presence of fertilizers, natural or placed by humans. On the other hand, denitrifying processes, which release N_2O through incomplete denitrification, are carried out by diverse bacterial communities, through genetically conserved denitrifying enzymes. According to a study in 2011, soil microorganisms react to warmer temperatures by differential expression of genes related to decomposition and nitrogen cycling (Zhou et al. 2011).

The increased levels of CO_2 in the environment is known to alter the release of nutrients from plant roots, which also leads to altering of microbial communities in the soil by stimulating growth, leading to a shift in CO_2 flux in the environment. A concern about this is that in the long term, the increased microbial biomass may cause immobilization of nitrogen in the soil, thereby removing it from circulation and decreasing the overall availability of it for plants. Also, this tipped ratio of carbon to nitrogen in the soil favors fungal dominance in the soil. Most species of fungi store carbon more than they metabolize, and their cell walls are more resistant to degradation. Therefore, in soil dominated by fungal species the rates of respiration are lower, potentially leading to carbon sequestration; in other words, long term storage of carbon. Several studies have reported that soil respiration rates increase in response to increased atmospheric CO_2 (Norby et al. 2004, Hungate et al. 1997), the effects more intense in below ground species (Jackson et al. 2009). In contrast to these results, some studies show that microorganisms that prefer labile carbons will instead metabolize those over complex carbons and organic matter decomposition will slow down, leading again to carbon sequestration.

Their huge impacts considered, it is no surprise that manipulation of microbial communities to mitigate anthropogenic climate change attracted researchers' attention over the last few decades. There is an urgent need to find novel ways to reduce greenhouse gas emissions and remove them from the atmosphere. Altering land usage to promote biotic carbon sequestration is a cost effective method with many benefits for the environment, such as better wildlife habitat and improved water and soil nutrient qualities, and this approach has received considerable attention in the last few decades (Busse et al. 2009, Lal 2008). However, its capacity for carbon removal is inefficient, in 25 to 50 years removing approximately 50–100 gigatons of carbon (GtC) (Lal et al. 2004), while global carbon budget 2018 estimated that anthropogenic activities release 9.4 GtC per year (Quere et al. 2018).

Energy

A great concern on the 21st century has been the growing demand for energy and the diminishing resources to provide it. According to data from International Energy Agency, fossil fuels currently make up more than 80% of the world's energy supply; however, fossil fuels are non-renewable, and crude oil, natural gas, and charcoal are being depleted, according to a projection by International Energy Agency; at this consumption rate, they will last for only 45, 60, and 120 years, respectively. As such, the need for developing novel methods and alternatives for energy supplies to replace fossil fuels is urgent. Bioenergy has been gathering much attention due to the wide availability of biomass to be used as feedstock. According to International Renewable Energy Agency, currently it makes up 10% of the total energy globally consumed, of which half is dependent upon biomass usage. While solid biomass for energy usage has been around for a long time in the form of firewood, recent advances in biotechnology have allowed the development of more efficient uses of firewood, wood chips, wood pellets and charcoal as solid biofuel.

Biofuels refer to raw biomass and its refined versions, converted to fuels in gas, solid or liquid forms. Biofuels have been at the center of many related discussions, providing a promising alternative to fossil fuels. Biofuels offer a promising alternative to reduce the greenhouse gas emissions as well as removing them from the atmosphere through the growth of fuel crops. There are four main parts in biorefineries: harvesting the feedstock, physicochemical/thermal pretreatment, enzymatic scarification and bioethanol conversion of sugars through fermentation. The current working bio-refinery utilize enzymatic and thermochemical treatments together. Most of the research has been towards lignocellulosic biomass as feedstock; however, more recent researches have also moved towards other options as feedstock, for example microorganisms (Adeniyi et al. 2018). Lignocellulosic biomass is mainly composed of three constituents: cellulose, lignin and hemicellulose, and they reside in a complex structure that is very hard to separate.

There are various types of liquid biofuels, which have gained attention due to their ease of transport. Some examples of these are bioethanol, biodiesel, methanol and biobutanol, all of which are alternatives to gasoline/ethane, diesel, methane and butane/gasoline, respectively. There are, however, challenges standing in the way of biofuels: bioethanol can only provide half the amount of energy per volume as gasoline, and while it produces less carbon monoxide, it releases more ozone. Methanol proves advantageous over methane in that it is liquid and easy to transport; however; it provides less than half energy per volume. Biodiesel is only slightly less efficient than regular diesel; however, it is corrosive, and therefore the engines have to be replaced for transition. First generation bioethanol was produced from sugarcane and corn; however, this was not sustainable as the need for land for the crops, as well as the growing of said crops drawing resources was not efficient. Ligninolytic enzymes produced mainly by fungi showed many properties fitting for an efficient delignification process. Currently, microbial engineering research for biofuel production aims to increase metabolic yields and product/inhibitor tolerance in addition to the development of novel methods for pretreatment in refineries. With the biotechnological developments, the research for microbial modifications for such processes is becoming easier, the advancements in synthetic biology even allowing for synthetic candidates that are designed for specifically this purpose from bottom up, therefore enabling more suitable alternative production methods and usage of already existing industrial transportation infrastructure.

In most bioprocesses that produce biofuels from lignocellulosic biomass, the main component of woody plants' cell wall structures, lignin is extracted in order to utilize the remaining cellulose. It currently remains a challenge to utilize lignin; it is the second most abundant biopolymer after cellulose, and constitutes up to 40% of the lignocellulosic biomass. While pulp and paper industry produced 50 million tons of extracted waste lignin in 2014, 98% of it was burned as low value fuel. Lignin degradation in nature is carried out by several fungal as well as some bacterial species, fungi being more efficient ones (Sigoillot et al. 2012), with over 1600 species that degrade lignin (Gilbertson 1980). The main bacterial species that degrade lignin consist of three species, α- and γ-Proteobacteria and Actinomycetes (Bugg et al. 2011a, Huang et al. 2013), and they are similar to fungi in several aspects including growth patterns, mycelia development and morphogenetic qualities. They also produce extracellular enzymes and utilize secondary metabolites for lignocellulosic matter decomposition. In this task, the most characterized ligninolytic enzymes are extracellular peroxidases and laccase, and genes for various forms of it have been identified in various archaea and bacterial species (Tian et al. 2014). While bacteria have recently been getting attention, saprotroph fungi still remain the primary degraders of lignin in nature (Huang et al. 2013). Saprotroph fungi can be categorized into three subspecies, white-rot fungi, brown-rot fungi and soft-rot fungi, with only white-rot fungi being able to completely degrade lignin into CO_2 and H_2O. The enzymes that degrade lignin can be investigated in two groups: lignin modifying enzymes, that take part directly in lignin degradation, and lignin degrading auxiliary enzymes, which cannot directly degrade lignin but are required for the processes. There are four most common applications of lignin modifying enzymes in delignification processes: fungal-based delignification, enzyme assisted delignification, laccase mediator systems and integrated fungal fermentation (Salvachúa et al. 2011, Wan et al. 2010).

Consortia of microorganisms rather than a single species have been reported to perform better (Mikesková et al. 2012). Studies have reported several combinations of cultures as more efficient novel methods, either as catalysts on pretreatment of lignocellulosic biomasses (Duan 2008) or pulping (Wang et al. 2013). With the fast-developing methods of biotechnology and screening, investigation of microbial consortia offer promising new platforms for lignocellulosic biomass degradation for various applications.

Microbial enzymes such as laccases or versatile peroxidases are focus of research for scientific communities that search for novel platforms for lignin degradation. The microbial enzymes vary depending on the requirements of the process; for example, one of the most studied ligninolytic bacterium, *Thermobifida fusca*, is an aerobic thermophilic bacterium belonging to actinomycetes, which produces thermo-tolerant extracellular cellulase and laccases, allowing for high temperature processes.

Aside from their uses in lignin degradation for processes such as bioethanol production, ligninolytic enzymes also show potential in removal of toxic chemical compounds from sugar hydrolysates (Placido et al. 2015).

References

Adeniyi, O. M., Azimov, U. and Burluka, A. 2018. Algae biofuel: Current status and future applications. Renewable and Sustainable Energy Reviews 90(March): 316–335. https://doi.org/10.1016/j.rser.2018.03.067.

Amann, M. 2008. Health Risks of Ozone from Long-Range Transboundary Air Pollution. Copenhagen: World Health Organization, Regional Office for Europe.

Archer, D. B. 2000. Filamentous fungi as microbial cell factories for food use. Current Opinion in Biotechnology 11(5): 478–483. Doi: 10.1016/s0958-1669(00)00129-4.

Aslan, S. and Kapdan, I. K. 2006. Batch kinetics of nitrogen and phosphorus removal from synthetic wastewater by algae. Ecological Engineering 28(1): 64–70. https://doi.org/10.1016/j.ecoleng.2006.04.003.

Bardgett, R. D. and van der Putten, W. H. 2014. Belowground biodiversity and ecosystem functioning. Nature 515(7528): 505–511. https://doi.org/10.1038/nature13855.

Bugg, T. D., Ahmad, M., Hardiman, E. M. and Rahmanpour 2011a. Pathways for degradation of lignin in bacteria and fungi. Natural Product Reports 28: 1883–96.

Buschke, N., Schäfer, R., Becker, J. and Wittmann, C. 2013. Metabolic engineering of industrial platform microorganisms for biorefinery applications—Optimization of substrate spectrum and process robustness by rational and evolutive strategies. Bioresource Technology 135: 544–554. https://doi.org/10.1016/j.biortech.2012.11.047.

Busse, M. D., Sanchez, F. G., Ratcliff, A. W., Butnor, J. R., Carter, E. A. and Powers, R. F. 2009. Soil carbon sequestration and changes in fungal and bacterial biomass following incorporation of forest residues. Soil Biology and Biochemistry 41(2): 220–227. Doi: 10.1016/j.soilbio.2008.10.012.

Cai, T., Park, S. Y. and Li, Y. 2013. Nutrient recovery from wastewater streams by microalgae: Status and prospects. Renewable and Sustainable Energy Reviews 19: 360–369. https://doi.org/10.1016/j.rser.2012.11.030.

Dalby, P. A. 2011. Strategy and success for the directed evolution of enzymes. Current Opinion in Structural Biology 21(4): 473–480. https://doi.org/10.1016/j.sbi.2011.05.003.

Editor, S. and Steinbu, A. 2011. Microbiology Monographs Series Editor: Alexander Steinbu. In Molecular Microbiology (Vol. 20). https://doi.org/10.1007/978-3-662-45209-7.

Fisher, A. K., Freedman, B. G., Bevan, D. R. and Senger, R. S. 2014. A review of metabolic and enzymatic engineering strategies for designing and optimizing performance of microbial cell factories. Computational and Structural Biotechnology Journal 11(18): 91–99. https://doi.org/10.1016/j.csbj.2014.08.010.

Gilbertson, R. L. 1980. Wood-rotting fungi of North-America. Mycologia 72: 1–49.

Gonçalves, A. L., Pires, J. C. M. and Simões, M. 2017. A review on the use of microalgal consortia for wastewater treatment. Algal Research 24: 403–415. https://doi.org/10.1016/j.algal.2016.11.008.

Heider, S. A. E. and Wendisch, V. F. 2015. Engineering microbial cell factories: Metabolic engineering of *Corynebacterium glutamicum* with a focus on non-natural products. Biotechnology Journal 10(8): 1170–1184. https://doi.org/10.1002/biot.201400590.

Hess, A., Zarda, B., Hahn, D., Häner, A., Stax, D., Höhener, P. et al. 1997. *In situ* analysis of denitrifying toluene- and m-xylene-degrading bacteria in a diesel fuel-contaminated laboratory aquifer column. Applied and Environmental Microbiology 63(6): 2136–2141.

Höhener, P., Hunkeler, D., Hess, A., Bregnard, T. and Zeyer, J. 1998. Methodology for the evaluation of engineered *in situ* bioremediation: Lessons from a case study. Journal of Microbiological Methods 32(2): 179–192. https://doi.org/10.1016/S0167-7012(98)00022-0.

Huang, X. F., Santhanam, N., Badri, D. V., Hunter, W., Manter, D., Decker, S. et al. 2013. Isolation and characterization of lignin-degrading bacteria from rainforest soils. Biotechnology and Bioengineering 110(6): 1616–1626. Doi: 10.1002/bit.24833.

Hungate, B. A., Holland, E. A., Jackson, R. B., Chapin, F. S., Mooney, H. A. and Field, C. B. 1997. The fate of carbon in grasslands under carbon dioxide enrichment. Nature 388(6642): 576–579. https://doi.org/10.1038/41550.

IEA, World energy outlook. 2013. International Energy Agency, Paris, France (2013).

Intergovernmental Panel on Climate Change. 2007. Climate Change 2007: the Physical Science Basis. Solomon, S. et al. (eds.). Contribution of Working Group I to the Fourth Assessment Report of the Intergovernmental Panel on Climate Change. Cambridge Univ. Press, Cambridge, UK.

Intergovernmental Panel on Climate Change. Global Warming of 1.5°C (IPCC, 2018).

Jackson, R., Cook, C., Pippen, J. and Palmer, S. 2009. Increased belowground biomass and soil CO_2 fluxes after a decade of carbon dioxide enrichment in a warm-temperate forest. Ecology 90(12): 3352–3366. doi: 10.1890/08-1609.1.

Jakočiunas, T., Jensen, M. K. and Keasling, J. D. 2016. CRISPR/Cas9 advances engineering of microbial cell factories. Metabolic Engineering 34: 44–59. https://doi.org/10.1016/j.ymben.2015.12.003.

Kalinowski, J., Bathe, B., Bartels, D., Bischoff, N., Bott, M. and Burko, A. 2003. The complete *Corynebacterium glutamicum* ATCC 13032 genome sequence and its impact on the production of L-aspartate derived amino acids and vitamins. 104. https://doi.org/10.1016/S0168-1656(03)00154-8.

Kamaludeen, S. P. B., Arunkumar, K. R., Avudainayagam, S. and Ramasamy, K. 2003. Bioremediation of chromium contaminated environments. Indian Journal of Experimental Biology 41(9): 972–985. Retrieved from http://www.ncbi.nlm.nih.gov/pubmed/15242290.

Kumar, A., Bisht, B., Joshi, V. and Dhewa, T. 2011. Review on bioremediation of polluted environment: A management tool. International Journal of Environmental Sciences 1(6): 1079–1093. https://doi.org/Article.

Lal, R. 2004. Agricultural activities and the global carbon cycle. Nutrient Cycling in Agroecosystems 70(2): 103–116. https://doi.org/10.1023/B:FRES.0000048480.24274.0f.

Lal, R. 2008. Carbon sequestration. Philosophical Transactions of the Royal Society B: Biological Sciences 363(1492): 815–830. doi: 10.1098/rstb.2007.2185.

Liu, Z., Si, B., Li, J., He, J., Zhang, C., Lu, Y. et al. 2018. Bioprocess engineering for biohythane production from low-grade waste biomass: technical challenges towards scale up. Current Opinion in Biotechnology 50: 25–31. https://doi.org/10.1016/j.copbio.2017.08.014.

Loera-Quezada, M. M., Leyva-González, M. A., Velázquez-Juárez, G., Sanchez-Calderón, L., Do Nascimento, M., López-Arredondo, D. et al. 2016. A novel genetic engineering platform for the effective management of biological contaminants for the production of microalgae. Plant Biotechnology Journal 14(10): 2066–2076. https://doi.org/10.1111/pbi.12564.

Muñoz, R. and Guieysse, B. 2006. Algal-bacterial processes for the treatment of hazardous contaminants: A review. Water Research 40(15): 2799–2815. https://doi.org/10.1016/j.watres.2006.06.011.

Nikhil, G. N., Venkata Mohan, S. and Swamy, Y. V. 2014. Systematic approach to assess biohydrogen potential of anaerobic sludge and soil rhizobia as biocatalysts: Influence of crucial factors affecting acidogenic fermentation. Bioresource Technology 165(C): 323–331. https://doi.org/10.1016/j.biortech.2014.02.097.

Norby, R., Ledford, J., Reilly, C., Miller, N. and O'Neill, E. 2004. Fine-root production dominates response of a deciduous forest to atmospheric CO_2 enrichment. Proceedings of the National Academy of Sciences 101(26): 9689–9693. doi: 10.1073/pnas.0403491101.

OECD. 2012. OECD Environmental Outlook to 2050: The Consequences of Inaction, OECD Publishing, Paris.

Otero, J. M. and Nielsen, J. 2010. Industrial systems biology. Biotechnology and Bioengineering 105(3): 439–460. https://doi.org/10.1002/bit.22592.

Pasupuleti, S. B. and Venkata Mohan, S. 2015. Single-stage fermentation process for high-value biohythane production with the treatment of distillery spent-wash. Bioresource Technology 189: 177–185. https://doi.org/10.1016/j.biortech.2015.03.128.

Plácido, J. and Capareda, S. 2015. Ligninolytic enzymes: A biotechnological alternative for bioethanol production. Bioresources and Bioprocessing 2(1). https://doi.org/10.1186/s40643-015-0049-5.

Quéré, C., Andrew, R., Friedlingstein, P., Sitch, S., Hauck, J., Pongratz, J. et al. 2018. Global carbon budget 2018. Earth System Science Data. https://doi.org/10.5194/essd-10-2141-2018.

Rawat, I., Ranjith Kumar, R., Mutanda, T. and Bux, F. 2011. Dual role of microalgae: Phycoremediation of domestic wastewater and biomass production for sustainable biofuels production. Applied Energy 88(10): 3411–3424. https://doi.org/10.1016/j.apenergy.2010.11.025.

Renuka, N., Sood, A., Ratha, S. K., Prasanna, R. and Ahluwalia, A. S. 2013. Evaluation of microalgal consortia for treatment of primary treated sewage effluent and biomass production. Journal of Applied Phycology 25(5): 1529–1537. https://doi.org/10.1007/s10811-013-9982-x.

Richey, A., Thomas, B., Lo, M., Reager, J., Famiglietti, J., Voss, K. et al. 2015. Quantifying renewable groundwater stress with GRACE. Water Resources Research 51(7): 5217–5238. doi: 10.1002/2015wr017349.

Salvachúa, D., Prieto, A., López-Abelairas, M., Lu-Chau, T., Martínez, Á. T. and Martínez, M. J. 2011. Fungal pretreatment: an alternative in second-generation ethanol from wheat straw. Bioresource Technology 102(16): 7500–6.

Sarkar, O., Kumar, A. N., Dahiya, S., Krishna, K. V., Yeruva, D. K. and Mohan, S. V. 2016. Regulation of acidogenic metabolism towards enhanced short chain fatty acid biosynthesis from waste: Metagenomic profiling. RSC Advances 6(22): 18641–18653. https://doi.org/10.1039/c5ra24254a.

Sharma, D. 2018. Microbial cell factories. Microbial Cell Factories 2–5. https://doi.org/10.1201/b22219.

Sigoillot, J. C., Berrin, J. G., Bey, M., Lesage-Meessen, L., Levasseur, A., Lomascolo, A. et al. 2012. Fungal strategies for lignin degradation. In Advances in Botanical Research (1st ed., Vol. 61). https://doi.org/10.1016/B978-0-12-416023-1.00008-2.

Smith, P., Fang, C., Dawson, J. J. C. and Moncrieff, J. B. 2008. Impact of global warming on soil organic carbon. Advances in Agronomy 97: 1–43. https://doi.org/10.1016/S0065-2113(07)00001-6.

Soberón-Chávez, G. 2011. Biosurfactants. Heidelberg: Springer-Verlag.

Strong, P. J. and Burgess, J. E. 2008. Treatment methods for wine-related and distillery wastewaters: A review. Bioremediation Journal 12(2): 70–87. https://doi.org/10.1080/10889860802060063.

Tian, J. H., Pourcher, A. M., Bouchez, T., Gelhaye, E. and Peu, P. 2014. Occurrence of lignin degradation genotypes and phenotypes among prokaryotes. Applied Microbiology and Biotechnology 98: 9527–44.

U.S. EPA. Report on the 2011 U.S. Environmental Protection Agency (EPA) Decontamination Research and Development Conference. U.S. Environmental Protection Agency, Washington, DC, EPA/600/R/12/557, 2012.

Venkata Mohan, S., Nikhil, G. N., Chiranjeevi, P., Nagendranatha Reddy, C., Rohit, M. V., Kumar, A. N. et al. 2016. Waste biorefinery models towards sustainable circular bioeconomy: Critical review and future perspectives. Bioresource Technology 215: 2–12. https://doi.org/10.1016/j.biortech.2016.03.130.

Vivaldi, M. 2011. Bioremediation—An overview. Journal of Industrial Pollution Control 27(2): 161–168. https://doi.org/10.1351/pac200173071163.

Wan, C. and Li, Y. 2010. Microbial delignification of corn stover by Ceriporiopsis subvermispora for improving cellulose digestibility. Enzyme and Microbial Technology 47(1-2): 31–36. Doi: 10.1016/j. enzmictec.2010.04.001.

Xie, S., Ragauskas, A. J. and Yuan, J. S. 2016. Lignin conversion: Opportunities and challenges for the integrated biorefinery. Industrial Biotechnology 12(3): 161–167. https://doi.org/10.1089/ ind.2016.0007.

Zhou, J., Deng, Y., Luo, F., He, Z. and Yang, Y. 2011. Phylogenetic molecular ecological network of soil microbial communities in response to elevated CO_2. mBio 2(4). Doi: 10.1128/mbio.00122-11.

CHAPTER 6

Composition and Biological Activities

Asmaa Missoum

Introduction

In 1928, Alexander Fleming discovered world's first antibiotic penicillin, which revolutionized the class of anti-microbial drugs. Since then further antibiotics were discovered to treat infectious diseases (Debbab et al. 2010). However, due to bacteria developing resistance, new discoveries have never ceased and seeking more effective therapeutic drugs from natural products has become a challenge in the medicinal field (Bérdy 2005). Research on chemistry of natural drugs is competitive with that of other synthetic products, as they are less toxic, biodegradable, and contribute to broad spectrum activities, even when administering lower quantities (Mossa et al. 2018). Nevertheless, research on natural products demands continuous enhancements in the screening process, which includes extraction, isolation, and metabolite structure interpretation. Such progress has provided the ability to conduct large number of bioassays, in high-throughput screening, with limited human intervention (Janardhan et al. 2014, Winn et al. 2018). This has also advocated a return to screening natural drug molecules, which continues to provide an important number of therapeutics approved for human use. Moreover, significant advances in synthetic biology, bioinformatics, and genomics can assist in altering natural products by biological and chemical means to boost their effectiveness (Selama et al. 2014).

Interestingly, *Streptomyces* sp. are the key source for broad variety of biologically active compound, as they account for around 60% of the antibiotics discovered in 1990 and most of agriculturally important compounds. Since they are able to synthesize many classes of secondary metabolites with diverse bioactivities, they are extensively used in research by academicians and industrialists (Raghava Rao et al. 2017). The biofunctional activities of these actinomycetes may find application as antiviral, anticancer, antibiotics, and immunosuppressants, as well as agricultural

Paris-sud université, UFR Sciences, 15 Rue Georges Clemenceau, 91400 Orsay, France.
Email: amissoum93@gmail.com

agents such as insecticides, herbicides, and plant hormones. Nevertheless, other species of bacteria (*Bacillus, Pseudomonas* and *Nocardiopsis* sp.) and fungi (*Penicillium, Aspergillus,* and *Bipolaris* sp.), whether soil-borne or endophytic, have also demonstrated similar effects from pure substances and extracts obtained from biomass or culture broth (Raekiansyah et al. 2017). Within this context, it would be nearly impossible to cover all topics related to biological activities' investigation of the discovered natural products. Hence, this chapter will focus on bacterial and fungal bioactive molecules as components of extracts, and not as pure chemical substances. The reader is also referred to detailed findings from very recent studies, as well as proposed mechanisms of action that contribute to these bioactivities.

Anti-microbial agents

Antibacterial agents

Bacterial genera with a high chance to detect compounds of interest as biosources are reported to be *Actinomycetes, Bacilli,* and *Pseudomonads* (Bérdy 2005). Recently, many pharmaceutically significant antibiotics have been identified from these bioresources: factumycin and tetrangomycin produced by *Streptomyces globosus* DK15 and *Streptomyces ederensis* ST13 isolated from forest soils located in Vietnam (Charousová et al. 2018), Bagremycins F and G produced by marine-derived *Streptomyces* sp. ZZ745 (Zhang et al. 2018), and Anthracimycin B collected from Deep-Sea *Streptomyces cyaneofuscatus* M-169 cultures (Rodríguez et al. 2018). Other *Streptomyces* species such as *Streptomyces levis,* isolated from soil sample of north India, were proven to be effective against Gram negative bacteria (Singh et al. 2018). Besides, non-proteinaceous and heat-stable bioactive components of *Streptomyces* sp. SBT343 such as azalomycin, actinoramide D, and azamerone, derived from marine sponges, inhibited staphylococcal biofilm formation without affecting bacterial viability. Nevertheless, these properties make them good candidates for developing future therapies against contact lens-mediated ocular infections (Balasubramanian et al. 2017).

Bacteriocin isolated from soil *Bacillus subtilis* GAS101 has been reported to have proteinaceous nature and was able to disrupt the bacterial cell membrane as its mechanism of action. These are produced by ribosomes and secreted by bacteria, mainly *Bacillus*, for self-defense (Sharma et al. 2018). Soil *Bacillus* species isolated from Tunisian arid areas also harbors high inhibitory potentials against clinical isolates of Gram negative bacteria. The identified active component was identified as 1-acetyl-β-carboline (Nasfi et al. 2018). Nevertheless, various other research groups including our university laboratory have also been responsible for screening studies. Many different *Bacillus* species, derived from Qatari coastal soils, have demonstrated significant antimicrobial activities against *Staphylococcus aureus, Staphylococcus epidermis,* and *Escherichia coli.* Unfortunately, the potent bioactive metabolites were not characterized in this study (Missoum and Al-Thani 2017). Nevertheless, significant biomolecules with antibacterial activities are presented in Fig. 1.

Most interestingly, different chemical compounds from marine-derived fungus *Aspergillus* sp. scs-kfd66 inhibited the growth of pathogenic bacteria such as *Escherichia coli, Staphylococcus aureus,* and *Listeria monocytogenes* (An et al.

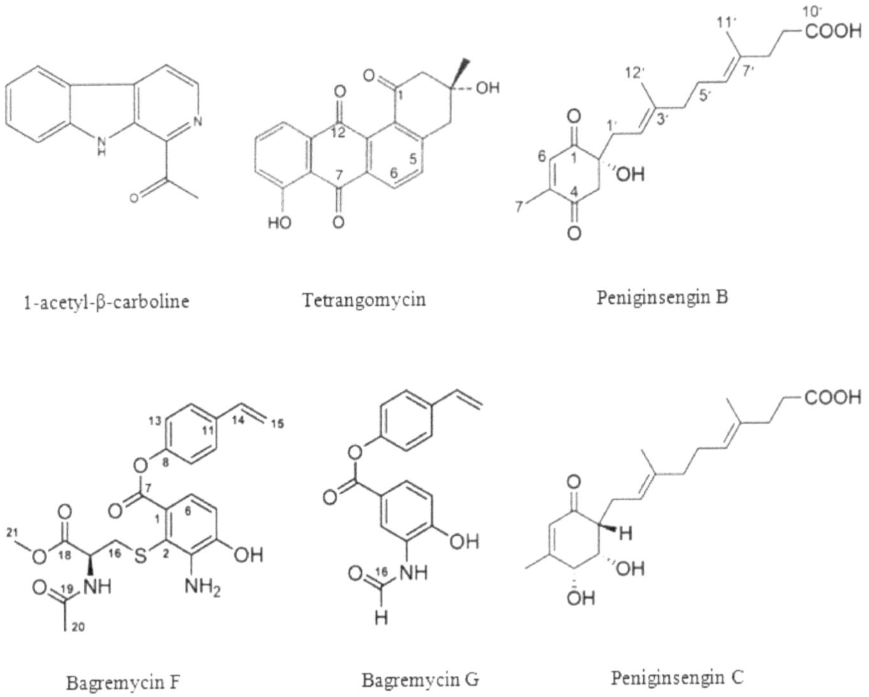

1-acetyl-β-carboline   Tetrangomycin   Peniginsengin B

Bagremycin F   Bagremycin G   Peniginsengin C

Fig. 1. Examples of antibacterial compounds and their structures (Nasfi et al. 2018, Charousová et al. 2018, Zhang et al. 2018, Cheng et al. 2018).

2018). The genus *Penicillium,* which is one of the most recognized genera of fungi for the discovery of bioactive constitutes, remains to provide major antibacterial drugs. In fact, newly discovered Farnesylcyclohexenones as Peniginsengins B–E, isolated from Deep Sea-derived *Penicillium* sp. YPGA11, exhibited similar promising effects (Cheng et al. 2018).

Antifungal agents

Fungal infections can be divided into two categories, systematic mycosis or dermatomycosis. The former is often lethal and has attracted more attention in recent research of mycology due to etiological changes, which mainly has occurred as a result of using broad spectrum antibacterial therapies. Consequently, opportunistic infections with fungi such as *Candida* sp. have increased gradually. These are also more prominent in immuno-compromised and intensive care unit patients (Srivastava and Dubey 2016, Córdova-Dávalos et al. 2018). Most of the important antimicrobial compounds including those having antifungal bioactivities are summarized in Table 1.

It was reported for the first time that uncharacterized biomolecules from *Streptomyces chrestomyceticus* ADP4 inhibited various *Candida albicans* strain at MIC_{90} values ranging from to 3.70 ± 1.8 to 10.10 ± 2.6 µg/mL. Other strains such as *Candida krusei, Candida tropicalis,* and *Candida paripsilosis* were also inhibited but had higher MIC_{90} values. Scanning electron microscopic (SEM) studies showed

Table 1. Summary of recently isolated bioactive compounds and their antimicrobial activities.

Bioactive compound	Microorganism/ producer	Source	Biological activities	Reference
Factumycin, Tetrangomycin	*Streptomyces globosus* DK15, *Streptomyces ederensis* ST13	Forest soils located in Vietnam	Antibacterial	(Charousová et al. 2018)
Bagremycins F and G	*Streptomyces* sp. ZZ745	Marine		(Zhang et al. 2018)
Anthracimycin B	*Streptomyces cyaneofuscatus* M-169	Deep-Sea		(Rodríguez et al. 2018)
Bacteriocin	*Bacillus subtilis* GAS101	Soil		(Sharma et al. 2018)
1-acetyl-β-carboline Bacillomycin and Fengycin	*Bacillus* species	Tunisian arid areas		(Nasfi et al. 2018)
Peniginsengins B–E	*Penicillium* sp. YPGA11	Deep-Sea		(Cheng et al. 2018)
Emericellipsin A	*Emericellopsis alkalina*	Alkaline soil near Zheltyr Lake, Russia	Antifungal	(Rogozhin et al. 2018)
P18 peptide	*Bacillus subtilis*		Antiviral	(Starosila et al. 2017)
Hydroxy marilone C	*Streptomyces badius*	Egyptian soils	Antiviral, anti-cancer	(El Sayed et al. 2016)
Butenolide	*Streptomyces* sp. Smu03	*Elephas maximus* intestines	Antiviral	(Li et al. 2018)
Brefeldin A	*Penicillium* sp. FKI-7127			(Raekiansyah et al. 2017)
Xanthones	*Aspergillus iizukae*	Coastal saline soil		(Kang et al. 2018)
Proteins and carbohydrates	*Trichoderma viride*	Cucumber rhizosphere soil		(Awad et al. 2018)
Polyketides	*Streptomyces* sp. USC-16018	Marine	Antiparasitic	(Buedenbender et al. 2018)
Cyclodepsipeptide WS9326A and annimycin	*Streptomyces asterosporus* DSM 41452			(Zhang et al. 2017a)
Gancidin W	*Streptomyces* SUK10	*Shorea ovalis* trees		(Zin et al. 2017)
10-acetyl trichoderonic acid A, hydroheptelidic acid, and 6'-acetoxy-piliformic acid	*Nectria pseudotrichia*	*Caesalpinia echinata*		(Costa et al. 2018)

that the metabolites extract inhibited biofilm formation by disturbing cell membrane as well as preventing fungal cells from adhering to polystyrene surface and their conversion to the hyphal status (Srivastava et al. 2016). Bioactive constitutes from *Streptomyces* species also showed anti-biofilm activity in *C. albican*. At 2 gL^{-1} concentration, *Streptomyces toxytricini* Fz94 isolated from Egyptian soils gave 92% inhibition compared to 90% by Ketoconazole after 120 min (Sheir and Hafez 2017), while bioactive extracts produced by soil *Streptomyces* sp. GCAL-25 were three times more effective than the control fungicide, amphotericin B (Córdova-Dávalos et al. 2018). Further studies are required to purify these specific compounds and elucidate their structures.

Highly stable AFP protein secreted from *Aspergillus giganteus* selectively inhibits the growth of filamentous fungi without affecting the mammalian cells' viability. Their γ-core motif is known to destabilize the integrity of plasma membranes. In a recent study, molecular dynamics simulations and NMR spectroscopy were used to characterize the dynamical behavior of AFP with fungal model membranes in solution. It was also concluded that AFP protein does not destroy the fungal membrane integrity by pore formation but covers its surface using a multistep mechanism (Utesch et al. 2018). Other fungal peptides such as emericellipsin A isolated from an extremophile, *Emericellopsis alkalina,* inhibited the growth of all *Candida* species as well as filamentous fungi *A. fumigatus* KBP F24 and *A. niger* ATCC 16404 at a concentration of 40 µg/per disc (Rogozhin et al. 2018).

Antiviral agents

The major targets for antiviral chemotherapy are herpes, influenza, and human immunodeficiency virus (HIV). To fight viral infections, clinical drugs should inhibit the virus replication at certain concentrations without harming the host cells. Although the success rate of antiviral therapy is still far away from that of antibacterial therapy, the situation is progressing. In recent years, a number of new biomolecules have been found to be effective against a variety of viruses (Fig. 2).

P18 is a new peptide that was recently isolated from the probiotic strain of *Bacillus subtilis*. This has resulted in complete inhibition of the influenza virus at 12.5 to 100 g/ml concentrations. P18 demonstrated no toxic effect in cytotoxicity studies involving Madin-Darby canine kidney (MDCK) cells (Starosila et al. 2017). Besides, hydroxy marilone C produced by *Streptomyces badius* isolated from Egyptian soil, protected 50% of the virus-infected MDCK cells against H1N1 cytopathogenicity (EC$_{50}$) by 33.25% for 80 µg/ml. However, the maximum cytotoxicity was at 27.9% with IC$_{50}$ value of 128.1 µg/ml (El Sayed et al. 2016). Butenolide produced by *Streptomyces* sp. Smu03 obtained from *Elephas maximus* intestines was also shown to be effective against oseltamivir-resistant influenza A virus strain A/PR/8/34 at early stages of infection. It is suggested that this molecule may have interfered with the fusogenic process of hemagglutinin (HA) of the virus, thus blocking its entry into host cells (Li et al. 2018).

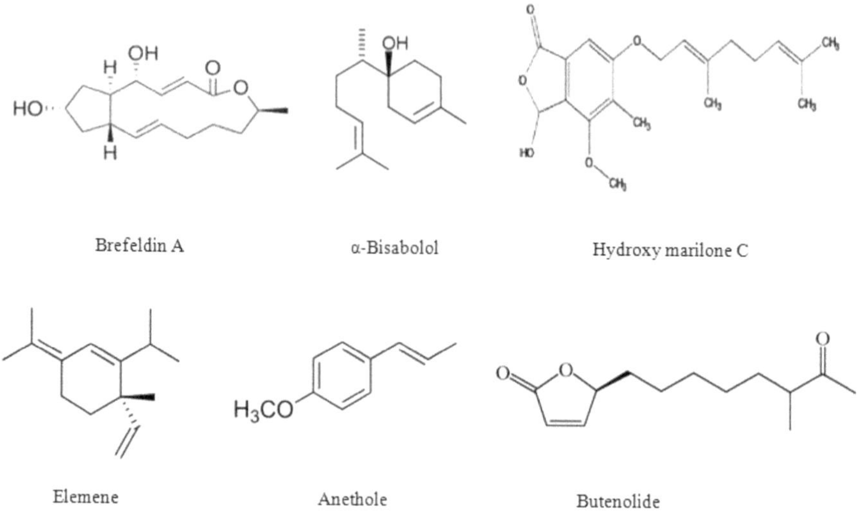

Brefeldin A   α-Bisabolol   Hydroxy marilone C

Elemene   Anethole   Butenolide

Fig. 2. Examples of antifungal and antiviral compounds and their structures (Raekiansyah et al. 2017, Li et al. 2018, El Sayed et al. 2016, Awad et al. 2018).

On the other hand, novel derived brefeldin A derived from *Penicillium* sp. FKI-7127 demonstrated significant inhibitory effects against different serotypes of dengue viruses as well as Zika and Japanese encephalitis viruses. The former is a mosquito-borne pathogen that causes dengue fever affecting 390 million people per year (Raekiansyah et al. 2017). Moreover, xanthones were extracted for the first time from *Aspergillus iizukae* fungus isolated from coastal saline soil. These compounds showed strong activities towards influenza virus H1N1, herpes simplex virus types 1 and 2 with respective IC_{50} values of 44.6, 21.4, and 76.7 μM (Kang et al. 2018). Proteins and carbohydrates from cucumber rhizosphere soil-derived fungus, *Trichoderma viride,* showed a moderate antiviral activity of 20% and 18%, respectively, against H5N1 virus at a concentration of 25 μg/μl (Awad et al. 2018).

Antiparasitic agents

Various parasitic diseases caused by protozoa and helminths invading the human body and domestic animals surely represent a major public health problem, particularly in tropical regions of the world. Among these diseases, malaria results in more than 400,000 deaths yearly. *Plasmodium* protozoan parasites that cause this vector-borne disease, use Anopheles mosquitos as the main host to spread it through human populations (Buedenbender et al. 2018). The emergence of parasites' resistance to currently commercially available drugs is one of the key factors that lead to infecting billions of people. Hence, new bioactive molecules are needed in order to reduce the risk of resistance and combat the spread of malaria as well as other parasitic diseases (Zin et al. 2017).

Polyketides isolated from marine-derived *Streptomyces* sp. USC-16018 demonstrated an inhibition of > 75% against *Plasmodium falciparum* strains, which was brought about through many mutations' types and have become resistant to nearly all antimalarial treatments including chloroquine (Buedenbender et al. 2018). Similarly, novel cyclodepsipeptide WS9326A and polyketide annimycin derivatives produced by *Streptomyces asterosporus* DSM 41452 exhibited modest inhibitory activity against *P. falciparum*, compared to its engineered mutant which yielded better results (Zhang et al. 2017). Gancidin W is another antimalarial agent isolated from *Streptomyces* SUK10, an endophytic to *Shorea ovalis* trees. This strain exhibited an *in vivo* inhibition rate of 80% against *Plasmodium berghei* PZZ1/100 with a very low toxicity (Zin et al. 2017).

Protozoan species of the genus *Leishmania* are also transmitted by the bites of infected female phlebotomine sandflies, affecting human populations in the Americas with an incidence rate of 19.76 cases per 100,000 inhabitants. Anhydrocochlioquinone A is a novel leishmanicidal that was recently isolated from the fungus *Cochliobolus* sp. Its ethyl acetate extract was shown to demonstrate 78% inhibition against *Leishmania amazonensis* amastigotes with an EC_{50} value of 22 lg/mL (42.7 lM) in the *in vitro* assay (Campos et al. 2017). Another endophytic fungus, *Nectria pseudotrichia*, isolated from *Caesalpinia echinata* has shown antiparasitical activity against *Leishmania (Viannia) braziliensis* with a potential low toxicity to THP-1 cells. Fractionation of the extract yielded 10-acetyl trichoderonic acid A, hydroheptelidic acid, and 6'-acetoxy-piliformic acid, which were the most active compounds with IC_{50} values of 21.4, 24.8, and 28.3 µM, respectively (Costa et al. 2018). Moreover, fungi species isolated from mangroves in Florida inhibited *Naegleria fowleri*, a brain-eating amoeba, by $> 67\%$ at 50 µg/mL (Demers et al. 2018).

Pharmacologically active substances

Antitumor agents

Each year, about 12.7 million new cases of cancer are reported worldwide in addition to resulted 7.6 million deaths. Current clinical treatments, which include chemotherapy, hormonal therapy, and immunotherapy, are usually associated with high toxicity. Other treatments such as radiotherapy and local surgeries may not be highly efficient due to poor accessibility to tumor site as targeting tumor cells is complicated. Other undesirable side effects such as non-specific cytotoxicity to normal cells increase the urgent demand for novel antitumor drugs, which improve cancer patient survival rate (El Sayed et al. 2016).

Natural products extracted from microorganisms have yielded effective drugs as they are capable of binding to three-dimensional biological receptors and complex proteins. Indeed, most bioactive compounds and their functional activities presented in this section are summarized in Table 2. Thus, they have also provided better insights into mechanisms of action. According to an *in vitro* study using MTT method, trichomide cyclodepsipeptides derived from marine fungus *Trichothecium roseum* has resulted in cytotoxic effects on different human cancer cell lines: SW480, human colon cancer; HL-60, human myeloid leukemia; and MCF-7, breast cancer; with IC_{50} values of 0.149, 0.107, and 0.079 µM, respectively (Zhou et al. 2018).

Endophytic fungi are also recognized as sources of pharmacologically important bioactive compounds. Recently, extracted compounds from seaweed

Table 2. Summary of recently isolated bioactive compounds and their pharmacological activities.

Bioactive compound	Microorganism/ producer	Source	Biological activities	Reference
Trichomide cyclodepsipeptides	*Trichothecium roseum*	Marine	Anti-cancer	(Zhou et al. 2018)
3-nitropropanoic acid, hexadecanoic acid, and octadecanoic acid	*Talaromyces purpureogenus*	Seaweed		(Kumari et al. 2018)
Breviones I	*Penicillium* sp. TJ403-1	Coral		(Yang et al. 2018)
Lipopeptaibol emericellipsin A	*Emericellopsis alkalina*	Alkalophilic soil		(Rogozhin et al. 2018)
Fungal immunomodulatory protein (FIP)	*Nectria haematococca*			(Xie et al. 2018)
Lobophorin K	*Streptomyces* sp. M-207	*Lophelia pertusa* coral collected from Avilés Canyon depths	Anti-cancer	(Braña et al. 2017)
Actinomycins V, D, and $X_{0\beta}$	*Streptomyces* sp. ZZ338	Marine		(Zhang et al. 2016)
Anthraquinone derivative	*Streptomyces* sp. ZZ406			(Chen et al. 2018a)
Diheptyl phthalate, maculosin, albocycline M-2, and dotriacontane	*Streptomyces sparsus* strain VSM-30	Deep sea marine samples of Bay of Bengal	Antioxidant	(Managamuri et al. 2017)
Chitosanase (CS038)	*Bacillus mycoides* TKU038	Taiwan soils	Anti-inflammatory	(Liang et al. 2016)
Streptochlorin	*Streptomyces* sp. 04DH110	Marine		(Shim et al. 2015)
Antimycin-type depsipeptide	*Streptomyces somaliensis* SCSIO ZH66	Deep sea		(Li et al. 2017)
Actinofuranones E-975 and E-492	*Streptomyces gramineus*	*Leptogium trichophorum* lichen		(Ma et al. 2018)
Curvularin-type polyketides	*Penicillium* sp. SF-5859	Marine		(Ha et al. 2017)
Mutolide	*Lepidosphaeria* sp. PM0651419	Horse dung samples collected from Rajkot, India		(Shah et al. 2015)

Table 2 contd. ...

... Table 2 contd.

Bioactive compound	Microorganism/ producer	Source	Biological activities	Reference
Lipoteichoic acid	*Streptomyces hygroscopicus* NRRL 2387T		Immuno-modulatory	(Cot et al. 2011)
ASK2	*Streptomyces* sp. ASK2	Rhizosphere soil of a medicinal plant		(Lalitha et al. 2017)
Nocapyrone H Nocardiopsins A and B	*Nocardiopsis* sp. KMF-001 *Nocardiopsis* sp. CMB-M0232	Marine		(Bennur et al. 2016)
Polysaccharides, L919/A and L919/B	*Lactobacillus casei* strain LOCK 0919	Fecal sample from a healthy 5-year-old boy		(Górska et al. 2016)
Streptopeptolin	*Streptomyces olivochromogenes* NBRC 3561		Chymotrypsin inhibitor	(Kodani et al. 2018)
Anthraquinone derivative	*Streptomyces* sp. ZZ406	*Anemone Haliplanella lineata*	Glioma glycolytic enzymes inhibitor	(Chen et al. 2018a)
Aspergifuranone	*Aspergillus* sp. 16-5B	*Sonneratia apetala* leaves	α-glucosidase inhibitor	(Liu et al. 2015)
Lasiodiplodins	*Trichoderma* sp. 307	*Clerodendrum inerme* stem bark	α-glucosidase inhibitor	(Zhang et al. 2017)
Sorokiniol	*Bipolaris sorokiniana* LK12	*Rhazya stricta*	Acetyl cholinestrase inhibitor	(Ali et al. 2016)

Talaromyces purpureogenus fungus such as 3-nitropropanoic acid, hexadecanoic acid, and octadecanoic acid were proven to be cytotoxic. MCF-7 and HeLa (cervical cancer) cells were the most affected by the ethyl acetate extract treatment, yielding respective IC_{50} values of 110 ± 3 and 101 ± 1.4 μg/mL. It was observed that induced cell membrane damage and mitochondrial depolarization has resulted in apoptosis in HeLa cells. However, the extract did not affect normal, HEK 293T, human embryonic kidney cell line (Kumari et al. 2018).

Similar findings were publicized by a study on breviones I, a new breviane spiroditerpenoid extracted from coral-derived fungus *Penicillium* sp. TJ403-1. It exhibited significant inhibitory activities against HL-60 (acute leukemia), HEP3B (hepatic cancer), and A-549 (lung cancer) cell lines with IC_{50} values of 4.92 ± 0.65, 5.50 ± 0.67, and 8.60 ± 1.36 μM, respectively (Yang et al. 2018). In another regard, Lipopeptaibol emericellipsin A peptide from alkalophilic, and soil fungus *Emericellopsis alkalina* exhibited a selective cytotoxic activity against HeLa and HepG2 cancer cell lines (EC_{50} < 0.5 and 2.8 μM, respectively), making it about 40 times less toxic to normal cells when compared with doxorubicin as a positive control. The mode of action of this short peptaibol is presumably associated with its effects on cell membranes (Rogozhin et al. 2018).

Isolated phleichrome from a phytopathogenic fungus, *Cladosporium Phlei,* was assessed as photosensitizers for photodynamic therapy (PDT). This new fungal perylenequinone displayed a good *in vitro*, photodynamic antitumor activity. MCF-7, HeLa, and SW480 cell lines had their lowest viability, at 10, 25, and 50 μg/mL treatment, respectively. However, all cell lines were more sensitive to phleichrome under illuminated conditions than under dark conditions. Discovering new photosensitizers is crucial as the currently used ones have several complications such as pain experienced during irradiation, skin sensitivity, and limited treatment depth (So et al. 2018).

Interestingly, fungal immunomodulatory protein, FIP, from *Nectria haematococca* was proven to negatively regulate PI3K/Akt signaling in lung adenocarcinoma cells, which induced cell cycle arrest, autophagy, and apoptosis. The G1/S and G2/M cell cycle arrest resulted from suppressed Akt phosphorylation, which contributed to the downregulation of CDK2, CDK4, cyclin B1 and cyclin D1, as well as the upregulation of p21 and p27 expression. In addition, autophagy and A549 apoptosis were promoted by decreasing mTOR phosphorylation along with increasing Bax/Bcl-2 and c-PARP expression ratio, respectively (Xie et al. 2018).

Other isolated haloalkaliphilic bacteria from Algerian Sahara desert soils were also screened for genes coding for antitumor compounds. It was found that isolate M5A, which was identified as *Actinopolyspora* sp., was positive for tryptophan dimer gene primers, showing 76% homology to the strain AR1455 gene cluster. The tryptophan dimerization enzymes were shown to biosynthesize structurally diverse tryptophan dimmers that in turn possess an antitumoral activity. *Streptomyces* sp. strain Ig6, on the other hand, showed a mixed PCR product that included a high intensity band for Glu1/Glu2 primers, corresponding to dNDP-glucose 4,6-dehydratase genes. These were previously used to search for anti-cancer biomolecules such as elloramycin, novobiocin, and ravidomycin (Selama et al. 2014). Furthermore, hydroxy marilone C, a secondary metabolite extracted from *Streptomyces badius,* was also shown effective against two tumor cell lines (MCF-7 and A-549). The calculated IC_{50} for the cell line A-549 was 443 μg/ml, indicating low anti-tumor affinity, while the other cell line MCF-7 demonstrated an IC_{50} of 147.9 μg/ml (El Sayed et al. 2016).

Lobophorin K is another anti-cancer metabolite that is isolated from marine *Streptomyces* sp. M-207, endophytic to cold-water coral *Lophelia pertusa* collected from Avilés Canyon depths. This new natural bioactive compound displayed cytotoxic activities against human pancreatic carcinoma (MiaPaca-2) and human breast adenocarcinoma (MCF-7) cell lines, with IC_{50} values of 34.0 ± 85.1 and 23.0 ± 8.9 μM respectively (Braña et al. 2017). Moreover, recently extracted secondary metabolites from *Streptomyces* sp. strain MUM265 contributed to $34.57 \pm 4.99\%$ reductions in cell viability of Caco-2 colon cancer cells at 400 μg/mL extract concentration. However, ED_{50}, the dose that is required to induce 50% cell viability reduction, was not achieved even at this highest concentration tested. Treatment with MUM265 strain extract caused abnormal rounding, cell shrinkage with reduced cytoplasm mass, and detachment of Caco-2 cells. It also induced potential depolarization of mitochondrial membrane, as well as DNA fragmentation and accumulation of subG$_1$ cells in cell cycle after 24 hours, suggesting potential apoptosis effects (Tan et al. 2019).

Other forms of cancer such as glicoma were also affected by marine *Streptomyces* sp. ZZ338 metabolites. The proliferation of rat glioma C$_6$ cells as well as human glioma U251 and SHG44 cells were inhibited with IC$_{50}$ values ranging from 0.42 to 1.80 nM for actinomycin V, 1.01 to 10.06 nM for actinomycin D, and 3.26 to 25.18 nM for actinomycin X$_{0\beta}$. In contrast, the control Doxorubicin, on the other hand, had IC$_{50}$ values ranging from 0.70 to 9.61 μM (Zhang et al. 2016). An anthraquinone derivative from *Streptomyces* sp. ZZ406 also demonstrated potent inhibition activity against the proliferation of human glioma SHG44, U251, and U87MG cells with IC$_{50}$ values ranging from 4.7 to 8.1 μM, compared to 1.9–9.6 μM of control Doxorubicin. In addition to having a high selectivity index (CC$_{50}$/IC$_{50}$, > 12.3 to 21.3), this compound also showed good stability in human liver microsomes (Chen et al. 2018).

Antioxidant agents

Antioxidants play an essential role in scavenging and inhibiting free radicals such as reactive oxygen species (ROS), whose increased levels result in oxidative stress and eventually induces damage to human cells by attacking DNA, proteins, and membrane lipids. ROS that include superoxide, hydroxyl, and peroxyl radicals can lead to severe health complications such as diabetes mellitus, neurodegenerative and inflammatory diseases, and even cancer. They have also been linked to deterioration of food products via lipid oxidation, which causes loss of essential fatty acids and fat-soluble vitamins. Although a number of synthetic antioxidants were used by pharmaceutical and food industries, they have been associated with potential health risks including carcinogenesis and liver damage (Tan et al. 2018). For this reason, modern research is now targeting naturally occurring antioxidative agents from microorganisms and plants that can serve as protective safe therapeutics. *Nocardiopsis* and *Streptomyces* species are two of the actinomycetes which gained enormous attention in pharmaceutical biotechnology, as recent studies focused on their antioxidant system responses under various oxidative stress conditions (Janardhan et al. 2014, Tan et al. 2018).

For instance, the potential isolate GN2 from Indian mangrove soil that was identified as *Nocardiopsis alba,* contained high phenolic content in its extracted secondary metabolites. This is considered as a key element for antioxidative efficiency as the two variables are positively correlated. The total antioxidant capacity was investigated in *in vitro,* yielding 2.72 ± 0.4, 2.95 ± 1.18, 3.05 ± 0.98, and 1.62 ± 0.4 AA/g for four different fractions, compared to 10.63 ± 0.85 AA/g for ascorbic acid as standard. The radical scavenging effect of fractions F2, F3, and F4 was higher than 50% at concentration of 50 μg/mL and at 100 μg/mL for F1 (Janardhan et al. 2014).

Other mangrove sites such as Kuala Selangor, Malaysia constitute a rich bioresource for *Streptomyces* bacteria. According to 16S rRNA phylogenetic analysis, isolated strain MUM292 showed gene sequence similarity of 99.54% with *Streptomyces griseoruber* NBRC12873. Its secondary metabolites have contributed to 35.98 ± 5.39% DPPH, 67.96±2.23% ABTS$^+$ radicals, and 79.23 ± 0.70% O_2^- scavenging activities, at a concentration of 4 mg/mL. Moreover, MUM292 extract also exhibited a significant metal-chelating activity of 22.54±2.37% as it reduced complex formation between Fe^{2+} ion and ferrozine, in addition to 28.94 ± 2.70% inhibition of iron-induced lipid peroxidation, at the same concentration (Tan et al. 2018). *Streptomyces* sp. MUM265 is another strain which demonstrated strong antioxidant capacities. Based on 16S rRNA gene sequence similarity, it was found to be closely related with *Streptomyces phaeoluteichromatogenes* NRRL 5799 and *Streptomyces misionensis* NBRC13603 by 99.8%. Strain MUM265 exhibited significant scavenging activities of 42.33±3.99% for DPPH, 88.50±0.37% for ABTS$^+$ radicals, and 55.99±1.03% for O_2^- radicals, at 4 mg/mL concentration. The ability of *Streptomyces* sp. MUM265 extract to chelate metals was also evaluated and the assay showed a 46.02±0.86% inhibition of ferrozine-Fe^{2+} complex formation at 4 mg/mL (Tan et al. 2019). The antioxidant potentials of *Streptomyces* strains are mainly attributed to the presence of cyclic dipeptides and phenolic compounds as bioactive agents as investigated using Pearson's correlation analysis (Tan et al. 2018, 2019).

Deep sea marine samples of Bay of Bengal are also a promising source for new bioactive compounds that remain widely unexplored. The ethyl acetate extract of isolated strain VSM-30, identified as *Streptomyces sparsus,* was subjected to chemotypic and GC-MS analysis. Findings revealed the presence of bioactive molecules known to possess antioxidant activities such as diheptyl phthalate, maculosin, albocycline M-2, and dotriacontane (Managamuri et al. 2017). Similarly, crude extract of isolated strain RD-5 from Gulf of Khambhat, which was the designated strain as *Streptomyces variabilis,* exhibited DPPH free radical scavenging and metal chelating activities of 82.86 and 89%, respectively, at 5.0 mg/mL, compared to 64% H_2O_2 scavenging activity at 0.05 mg/mL concentration. On the other hand, ascorbic acid used as standard control showed 74.5% H_2O_2 scavenging activity and 86% DPPH free radical scavenging activity at 0.05 mg/mL concentration (Dholakiya et al. 2017).

Anti-inflammatory

Inflammation is an important part of the human body's immune response, which reacts to harmful stimuli like endotoxin exposure, damaged cells, or microbial infections. Inflammatory signals either initiate and maintain the process or cease it, leading to damaged tissues and cells in case of asymmetry of these two signals. Macrophages are immune cells that play a critical role in inflammation and their stimulation by lipopolysaccharide (LPS) induces a range of enzymes and pro-inflammatory cytokines such as interleukins (ILs), cyclooxygenase-2 (COX-2), tumor necrosis factor-α (TNF-α), and inducible nitric oxide synthase (iNOS). These are involved in initiating several biochemical pathways in inflammatory conditions such as autoimmune diseases and rheumatoid arthritis (Ha et al. 2017). For example, LPS can bind CD14/TLR4/MD2 receptor complex in macrophages as well as B cells and monocytes, resulting in activation of protein kinases with MyD88 and adaptor protein TRAF6 recruitment. This leads to subsequent translocation of NF-κB into the nucleus, wherein it interacts with promoter regions of various genes encoding pro-inflammatory mediators (Shah et al. 2015).

A novel chitosanase, designated as CS038, was retrieved from *Bacillus mycoides* TKU038 sampled from a Taiwan soil. CS038 was shown to hamper nitric oxide (NO) production in *Escherichia coli* LPS-induced murine RAW 264.7 macrophages. CS038 chitooligomers with low degrees of polymerization possessed higher anti-inflammatory capacities with an IC_{50} value of 76.27 ± 1.49 µg/mL (Liang et al. 2016). On the other hand, *Streptomyces* species are an economically important group as they provide wider range of metabolites with potent anti-inflammatory effects. Streptochlorin, which is isolated from marine actinomycetes, was proven to attenuate production of pro-inflammatory mediators in LPS-stimulated RAW264.7 cells via inhibition of TRIF-dependent signaling pathways. These included NO, COX-2, IL-1β, and IL-6. In LPS-induced acute lung injury mouse model, streptochlorin suppressed the neutrophils infiltration into lungs, as well as TNF-α and IL-6 production in broncho-alveolar lavage fluid (Shim et al. 2015).

Using RBC membrane stabilization method, the anti-inflammatory activity of three bioactive compounds, extracted from mangrove soil-derived *Streptomyces coelicoflavus* BC 01, was investigated and expressed as % of inhibition. When compared to control Diclofenac sodium, the % of inhibition were found to be 82.86 ± 12.47 (BC 01_C3), 73.89 ± 12.50 (BC 01_C1), and 71.26 ± 2.53 (BC 01_C2), all at 20 µg/ml as the highest concentration. It was hypothesized that the three compounds could have hindered the extracellular release of neutrophils' lysosomal content at inflammation site, which causes further tissue damage due to proteases and bactericidal enzymes (Raghava Rao et al. 2017). Furthermore, an antimycin-type depsipeptide, isolated from deep-sea *Streptomyces somaliensis* SCSIO ZH66, significantly restrained IL-5 production with IC_{50} value of 0.57 µM, in ovalbumin-stimulated splenocytes at 1 µM concentration (Li et al. 2017). Likewise, two actinofuranones along with E-975, and E-492, derived from *Streptomyces gramineus* associated to *Leptogium trichophorum* lichen, attenuated NO production through suppression of iNOS expression in LPS-induced RAW 264.7 macrophage cells. They have as well managed to suppress LPS-stimulated release of IL-6 and TNF-α in a dose-related manner. It is worth mentioning that the anti-inflammatory capacity was noticeably influenced by the positioning of hydroxyl group at C-5 unsaturated alkyl chains. Less hydroxyl substitutes located at alkyl chain contributed to stronger anti-inflammation effect (Ma et al. 2018).

Fungal metabolites have also been proven to be potential anti-inflammatory agents. Curvularin-type polyketides from Marine-derived *Penicillium* sp. SF-5859 attenuated the expression of iNOS and COX-2 in LPS-stimulated RAW264.7 macrophages. One of the compounds, (10*E*,15*S*)-10,11-dehydrocurvularin, inhibited the nuclear factor-κB (NF-κB) signaling pathway, thus suppressing the upregulation of cytokines and pro-inflammatory mediators. Others strongly restrained LPS-induced overproduction of nitric oxide and prostaglandin E_2 with IC_{50} values of 1.9–18.1 µM and 2.8–18.7 µM, respectively (Ha et al. 2017). Furthermore, mutolide, extracted from coprophilous fungus *Lepidosphaeria* sp. PM0651419, also mitigated LPS-induced secretion of TNF-α and IL-6 from THP-1 cells, as it hindered cytokine IL-17 secretion from anti-hCD28/anti-hCD3 stimulated human peripheral blood mononuclear cells (hPBMCs). The activation of NF-κB, which is a major transcription factor implicated in the secretion of these pro-inflammatory cytokines, was inhibited by mutolide in a dose-dependent manner (Shah et al. 2015).

Immunomodulators

Immunological defense system involves a complex interaction between humoral and cellular immune responses, induction and repression of immunocompetent cells, as well as the influence of endocrine and other mechanisms. The functioning immunity is also mostly attributed to the involvement of a variety of lymphoid cells in the clearance of viruses, bacteria, and fungi. These microbial infections are linked to immunoregulation since improved cell-mediated or antibody immune responses could occur. However, since human immunity declines with old age, many diseases such as diabetes, nephrosis, arthritis, and even cancer are more likely to develop. Thus, immunomodulatory agents are required to treat immunodeficiency diseases and to be used as immuno-suppressors in order to reduce rejection of organ transplantation, in addition to treat autoimmune diseases such as dermatomycosis, rheumatoid arthritis, and systemic lupus erythematosus (Zimmermann et al. 2018).

It has been long known that many of the immunomodulatory agents that improve the immune response are derived from microorganisms. These are considered to be potent tools for studying biochemical and cellular events of the immune responses. For example, lipoteichoic acid isolated from *Streptomyces hygroscopicus* NRRL 2387T stimulated IL-6 and TNF-α production by HEK293 human macrophage cells. These were activated *via* recognition by human immune Toll-like receptor 2 (TLR2), which confirms its role in detecting invading microbes such as Gram positive bacteria. However, this activity depended partially on CD14, TLR1, and TLR6 (Cot et al. 2011). ASK2 is another *Streptomyces*-derived bioactive compound, which inhibited the Gram negative *Klebsiella pneumoniae* through opsonophagocytosis and macrophage cytokines modulation. Using 8 μg/ml of ASK2, the phagocytic response improved from 32 to 62% for RAW264.7 cells, and from 26 to 52% for J774.A.1 cells. This has resulted in elevated levels of NO generation and influenced the stimulation of IL-12, TNF-α, and IFN-γ proinflammatory cytokines, implying the importance of ASK2 opsonic role against multi-drug resistant *K. pneumoniae* (Lalitha et al. 2017). In addition, secondary metabolites from *Streptomyces calvus* reduced immunosuppressive cytokine IL-10 levels with increased IL-2 and IFN-γ gene expression in human PBMCs. In a dose-dependent manner, the proliferation of PBMCs significantly improved in response to metabolite treatment (Mahmoudi et al. 2016).

Interestingly, a recent systematic review reported that *Streptomyces* macrolides had various immunomodulatory capacities in different diseases such as blepharitis, asthma, periodontitis, cystic fibrosis, and chronic rhinosinusitis. Most studies found out that neutrophil function was inhibited more than eosinophil function, and that decline in T helper (Th) 1 cytokines (IL-2, INF-γ) was reported less frequently than a decline in Th 2 cells cytokines (IL-4, IL-5, IL-6). There was also recurrent decrease in neutrophils number as well as in IL-1β, IL-6, IL-8, TNF-α, eosinophilic cationic protein, neutrophil elastase, and matrix metalloproteinase 9 concentrations, as is summarized in Fig. 3 (Zimmermann et al. 2018). Another review targeting immunomodulatory compounds obtained from *Nocardiopsis* species has highlighted certain metabolites' activities such as Nocapyrone H, Nocardiopsins A and B. The first compound is a polyketide α-pyrones that was isolated from *Nocardiopsis* sp. KMF-001. It was reported to reduce brain inflammation through restraining the activation of immune effector microglia cells, which in turn reduced pro-inflammatory factors such as Prostaglandin 2 (PGE2) and IL-1β and inhibiting NO production. Besides, Nocardiopsins A and B are recently isolated polyketides from marine *Nocardiopsis* species (CMB-M0232), which exhibit immunosuppressive effect by binding to FKBP12, a cytosolic protein that belongs to the immunophilin family (Bennur et al. 2016).

Alternatively, the *Lactobacillus casei* strain LOCK 0919 also produced immunomodulatory polysaccharides, L919/A and L919/B, against third-party antigens. The use of probiotics with such properties is an interesting strategy for solving gut microflora imbalance. Thus, this strain from *Lactobacillus* genus was investigated as they are of particular interest, since they are aimed for the dietary management of atopic dermatitis and food allergies. It was concluded that both L919/A and L919/B altered the immune response of mouse BM-DC to *Lactobacillus planatarum* WCFS1, but only L919/B modulated the response of THP-1 cells in

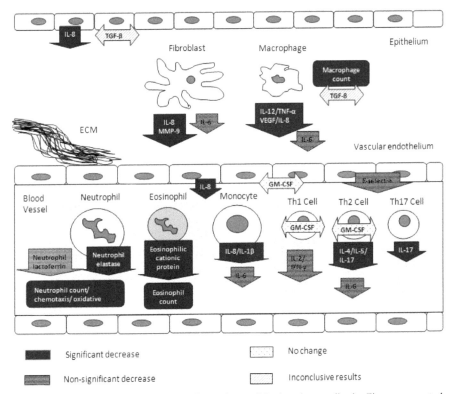

Fig. 3. Overview of immunomodulatory effects of macrolides based on studies by Zimmermann et al. (2018). Arrows depict excreted proteins; boxes depict cell counts or functions.

terms of TNF-α production. Additionally, TLR2, TLR4 and NOD2 were not involved in the recognition of the two polysaccharides (Górska et al. 2016).

Endophytic fungi are also a promising source for immunomodulatory new drugs. Secondary metabolites were extracted using ethyl acetate from the leaves and stems of *Newbouldia laevis, Agerantum conyzoides, Azadirachta indica,* and *Psidium guajava.* About 58.33% of the extracts demonstrated a significant counteracting activity against cyclophosphamide-induced reduction in percentage of total WBCs and monocyte. There was also a significant inhibition of heamagglutination and DTHR (Delayed Type Hypersensitivity Response) induced by SRBC in rats equal to 69.23 and 78.49%, respectively. According to HPLC-DAD analysis, luteolin, peperine, ferulic acid, epicatechin, rocaglamid, and phenylacetic acid were present in the extracts suggesting their role in this bioactivity (Treasure et al. 2017).

Enzyme inhibition

In the past decades, it has been considered that various enzymes have essential roles in the maintenance of homeostasis and that diseases result from its collapse. Some enzyme inhibitors are crucial pharmacologically active drugs since the correlations between disease processes and biological function of enzymes have been established. In addition, they are useful in elucidating the regulatory and biochemical pathways in the homeostasis of living organisms. Many chemically synthesized antagonists, which were subjected to structure-activity analysis, were used as drug until 1970s when the search for natural enzyme inhibitors had begun (Abdel-Shakour et al. 2015).

Recently, Shahzad et al. (2018) had reported that *Bacillus amyloliquefaciens* RWL-1, an endophytic to *Oryza sativa* seeds, produced α-glucosidase and urase inhibitor that exhibited $52.98 \pm 0.8\%$ and $51.27 \pm 1.0\%$ activities, compared to positive control values of $79.14 \pm 1.9\%$ and $88.24 \pm 2.2\%$, respectively. However, research on *streptomyces* species, once again, provided more bioactive metabolites as enzyme inhibitors. El-Hadedy and his workers have isolated and purified a protease inhibitor (STI1) from *Streptomyces lavendulae*, which inhibited Coxsackie B viral replication by 54%. Such findings are important for the development of antiviral drugs to treat HIV/AIDS and hepatitis C virus infections (El-Hadedy et al. 2015). Nevertheless, Kodani and his researchers isolated for the first time Streptopeptolin from *Streptomyces olivochromogenes* NBRC 3561 and demonstrated its inhibitory activity against chymotrypsin as 5.0 μg/mL IC_{50} value. This molecule's structure is related to cyanobacterial cyanopeptolin-type peptides, which were proven to inhibit serine proteases (Kodani et al. 2018).

Other studies focused more on searching for enzyme inhibitors from marine actinomycetes, since their characteristics are completely different from those of terrestrial actinobacteria. Extract of *Streptomyces* sp. Nyr04, isolated from Red Sea mangrove sediments, showed 76.25% α-amylase inhibition activity at a concentration of 60 mg/ml (Abdel-Shakour et al. 2015). In the screening work by Siddharth and Vittal's groups, volatile metabolites of *Streptomyces* sp. S2A collected from Indian coast sediments, showed inhibitory potential against α-amylase and α-glucosidase with IC_{50} values of 20.46 and 21.17 μg/mL, respectively (Siddharth and Vittal 2018). Furthermore, Kumar and Rao (2018) reported the isolation of α-glucosidase inhibitor from chloroform extract of *Streptomyces coelicoflavus* SRBVIT13, derived from Andhra Pradesh coastal soils. Oral treatment with 600 mg/kg of the extract has also resulted in a significant regulation of postprandial hyperglycaemia by 82.25% (Maltose) and 77.25% (sucrose) reduction in streptozotocin induced diabetic Wister rats.

Newly discovered anti-glioma agents from marine sources could serve as alternative treatment to current chemotherapies such as temozolomide (TMZ). This compound was used to treat malignant forms of human brain tumors (gliomas), but TMZ has low efficacy. The mechanism of Actinomycin D, derived from marine *Streptomyces* sp. ZZ338 of sea squirts, was reported for the first time. The treatment of U87-MG cells for 48 h with 0.01 nM Actinomycin D was found to downregulate multiple glioma metabolic enzymes such as PKM2 and HK2 from glycolysis, FASN from lipogenesis, and GLS from glutaminolysis (Zhang et al. 2016). Following this, Chen and his coworkers discovered novel anti-glicoma agents from another marine actinomycete strain, *Streptomyces* sp. ZZ406 derived from anemone *Haliplanella lineata*. Isolated compound 1, which was characterized as an anthraquinone derivative, was able to downregulate the expression of various glioma glycolytic enzymes in U87MG cells at 30.0 µM concentration. These included HK2, LDH, PKM2, and PFKFB (Chen et al. 2018a).

Endophytic fungi have also been investigated as a rich source for bioactive agents as enzyme inhibitors. Studies on mangrove-derived *Aspergillus* sp. 16-5B, an endophyte to *Sonneratia apetala* leaves, had led to the extraction and characterization of aspergifuranone. This metabolite showed a significant inhibitory activity against α-glucosidase with IC_{50} value of 9.05 ± 0.60 µM, compared to 553.7 ± 6.8 µM of acarbose as positive control. Kinetic analysis further confirmed the noncompetitive nature of this activity (Liu et al. 2015). Other α-glucosidase inhibitors include two characterized lasiodiplodins from *Trichoderma* sp. 307, an endophyte derived from *Clerodendrum inerme* stem bark. The active compounds were produced after co-cultivation with *Acinetobacter johnsonii* B2, and showed inhibitory activities with IC_{50} values of 25.8 and 54.6 µM, compared to 703.8 µM of acarbose as positive control (Zhang et al. 2017). Alternatively, sorokiniol from *Bipolaris sorokiniana* LK12 was reported for the first time to remarkably inhibit acetyl cholinestrase (EC_{50} value of 3.402 + 0.08 µg/mL). This bioactive compound from *Rhazya stricta* fungal endophyte could serve as a new therapeutic for Alzheimer's disease, in which cholinestrase inhibition is a major target (Ali et al. 2016).

Agrochemicals

Many economically valuable agricultural crops face biotic challenges such as phytopathogens. These cause devastating plant diseases that lead to the loss of crops. For this reason, there is an increasingly growing demand for the development of new fertilizers to inhibit or slow the growth of pathogen (Ben Mefteh 2018). However, current chemically synthesized biocontrol products that are available in the market may present serious side effects on both plant and human health. Therefore, it is desirable to search for natural sources of biocontrol products such as herbicidal and fungicidal compounds extracted from bacteria and fungi. Some may even contain plant growth-promoting factors that further helps in supporting, protecting, and maintaining plant growth (Patel et al. 2019).

Fungicides and bactericides

Among the recent findings on this topic, endophytic bacterium *Pseudomonas protegens* MP12 exhibited broad-spectrum antifungal activities against grapevine phytopathogens including *Alternaria alternata, Aspergillus niger, Botrytis cinerea, Neofusicoccum parvum* and *Penicillium expansum.* Compounds 2,4-diacetylphloroglucinol (2,4-DAPG), pyoluteorin, and pyrrolnitrin were considered the most effective, especially against *Phaeoacremonium aleophilum* and *Phaeomoniella chlamydospora,* which causes destructive tracheomycosis disease of grapevine trunks (Andreolli et al. 2019). Known derivatives of fungicides such as bacillomycin and fengycin were also isolated from soil *Bacillus* sp. (Nasfi et al. 2018). Moreover, sugarcane rhizosphere bacteria belonging to *Proteobacteria, Firmicutes,* and *Bacteroides*, were shown to be antagonistic to pathogenic *Colletotrichum falcatum* cfNAV, cfCHA, and cf8436, eventually, suppressing the red hot disease (Patel et al. 2019).

Actinomycete bacteria also contribute as biosources for natural fungicides. Strain CB-75, which was identified as *Streptomyces spectabilis* NBRC 13424 due to 99.93% sequence similarity, exhibited fungicidal capacity against various plant pathogenic fungi. Its ethyl acetate extract significantly inhibited spore germination at the concentration 2 × MIC and the mycelial radial growth of the following: *Colletotrichum musae* (80.96 ± 0.78%), *Colletotrichum acutatum* (79.85 ± 0.90%), *Alternaria tenuissima* (73.85 ± 0.84%), and *Fusarium oxysporum* (73.11 ± 0.80%). This broad-spectrum antifungal activity was attributable to detected non-ribosomal peptide synthetase (NRPS) and Type I polyketide synthase (PKS-I) biosynthetic genes. Pot experiments further proved that the treatment of banana seedlings with *Streptomyces* sp. CB-75 extract resulted in 83.12% prevention and control effect on banana *Fusarium* wilt. This strain was isolated from Chinese soils of diseased banana plantation, indicating it may have played a beneficial, rhizospheric role (Chen et al. 2018b).

Genera of endophytic fungi comprising of *Alternaria, Fusarium, Nigrospora, Pestalotiopsis, Phoma,* and *Xylaria* exhibit important antagonism activities against soil-borne fungus, *Fusarium solani,* a common devastating tree pathogen. These were isolated from *Rhizophora mucronata* marine mangrove species (Hamzah et al. 2018). Other sources of useful endophytes fungi include roots of date palm trees, *Phoenix dactylifera* L., which provided *Geotrichum candidum* TDPEF20 and *Penicillium citrinum* TDPEF34. These proved very effective against three pathogenic fungi *Trichoderma* sp. *(Ti), Trichoderma* sp. *(Ts),* and *Fusarium sporotrichioides (Fs),* in confrontation assays (Ben Mefteh 2018). Moreover, *Trichoderma viride* is widely recognized for its antagonistic ability towards vegetational pathogenic fungi. As biocontrol agent, it has gained a considerable attention as a novel study proved that it exhibited significant antifungal activities against *Fusarium solani, Fusarium oxysporium, Rhizoctonia solani, Pythium ultimum,* and *Sclerotium rolfsii* at 100 µg/disc (Awad et al. 2018).

Bactericidal compounds from *Streptomyces* species were able to inhibit *Burkholderia glumae*, a major causative agent *for* Bacterial Panicle Blight of rice. *Streptomyces* strains A20 and 5.1 significantly inhibited growths of a wide range of fungal and bacterial species (*Burkholderia gladioli, Acidovorax avenae, Klebsiella pneumonie,* and *Chromobacterium violaceum*), while strain 7.1 showed only antifungal activities (*Gaeumannomyces* sp., *Ulocladium* sp., *Colletotrichum* sp., and *Fusarium oxysporum*). Isolate A20, which was characterized as *Streptomyces racemochromogenes* DSM 40194 by sharing 97.2% similarity, possessed Streptotricins D, E and F as metabolites in its extract (Suárez-Moreno et al. 2019). Germination Arrest Factor (GAF) isolated from *Pseudomonas fluorescens* strain WH6 has been also shown to express specific biocidal activity against *Erwinia amylovora,* a bacterium that infects orchards and causes destructive fire blight disease (Harding and Raizada 2015). Other bacteriocides include sphaeropsidin A isolated from *Diplodia cupressi*, which showed a good activity against rice pathogens *Burkholderia glumae, Pseudomonas fuscovaginae,* and *X. oryzae* pv. *oryza* (Massi et al. 2018).

Herbicides

There are 6000 species of weeds, invasive plant species, affecting the agricultural production worldwide. Weeding is often carried out using chemically synthesized herbicides which either inhibit photosynthetic electron transfer, affect cell division and hormonal actions, or interfere with nutritive metabolism. However, many weeds developed resistance to existing herbicides that were proven to cause severe health issues in humans, and harmful pollution to the environment (Harding and Raizada 2015). Therefore, new natural herbicides are discovered and isolated from microorganism as bioactive compounds. These must be potentially biodegradable, minimally toxic to other living things, and highly selective against target weeds. Moreover, investigating and designing herbicides from such sources also allows studying action mechanism pathways of plant-specific enzymes inhibitors at molecular level (Masi et al. 2018, Winn et al. 2018).

Some of the typical herbicidal microbes are *Colletotrichum gloeosporioides* and *Colletotrichum orbiculare,* in which their metabolites include disease effectors such as small secreted proteins (SSPs) and cell wall degrading enzymes. Photobleaching macrocidins from *Phoma macrostoma* specifically target dicot plants by affecting new growth, which suggests that these compounds are transported in the phloem. *Phoma chenopodicola* also produces chenopodolin, a phytotoxic diterpene that resulted in necrotic lesions on creeping thistle (*Cirsium arvense*), annual mercury (*Mercurialis annua*), and green foxtail (*Setaria viridis*). Moreover, oxalic acid production by *Sclerotinia sclerotiorum* and *Sclerotinia minor* has effectively controlled dandelions and creeping thistle by acidifying host tissue, as well as inducing cell wall degradation and interfering with polyphenol oxidase (PPO) (Harding and Raizada 2015). Furthermore, penicillic acid is also recognized as a vital phytotoxin produced by *Penicillium canescens* and *Penicillium cyclopium,* that inhibits the germination of *Zea may* and alters the overall turnover of its metabolites (Bazioli et al. 2017).

In a recent review, various herbicidal activities of fungal phytotoxins were described in detail. Among the reported biocontrol agents, phomentrioloxin, isolated from *Phomopsis* sp., was used to control *Carthamus lanatus*; cytochalasans and nonenolides produced by *Ascochyta*, *Phoma*, and *Stagonospora* spp. for inhibiting *Carthamus arvense* and *Sonchus arvensis*; papyracillic acid and agropyrenol, isolated from *Ascochyta agropyrina* var. *nana* for controling *E. repens*; as well as chenopodolin derived from *Phoma chenopodiicola* for weeding *Cirsium arvense*. Cochliotoxin, radicinin, and 3-*epi*-radicinin, isolated from liquid cultures of *Cochliobolus australiensis,* were also shown to be effective against *Pennisetum ciliare* (Buffelgrass), according to leaf puncture and coleoptile elongation bioassay. Interestingly, sphaeropsidone, produced by *Diplodia cupressi*, exhibited the ability to stimulate haustorium development in radicals of *Orobanche crenata*, *Orobanche cumana*, and *Striga hermonthica,* in a dose-dependent manner. The haustorium is a plant organ in which parasitic weeds are used to withdraw plant nutrients after invading the host. Therefore, haustorium-inducing activity might be a suitable strategy to control these weeds in the absence of the host (Massi et al. 2018).

A number of bacterial herbicides have also been investigated as biocontrol agents, as they are characterized by rapid growth, suitability for genetic modification, and simple propagation requirements. Some of these include *Pseudomonas fluorescens* and *Xanthomonas campestris*. Strain D7 of the former species is originally isolated from the rhizospheres of *Triticum aestivum*. Analyses of its cell-free filtrates reveal presence of a lipopolysaccharide and extracellular peptides, which inhibit germination and growth of *Bromus tectoru*. On the contrary, strain WH6 affects germination of both monocot and dicot species via Germination Arrest Factor (GAF). This compound belongs to oxyvinylglycines class and has been shown to interfere with nitrogen metabolism and ethylene biosynthesis enzymes. Another strain of *P. fluorescens* species, BRG100, produces pseudophomins A and B as herbicidal agents against *Setaria viridis* (Harding and Raizada 2015).

Many biological agents do not have the essential bioactivity at the outset. Thus, further analogue synthesis and derivatization is required to develop optimized, efficient compounds. This optimization can be achieved through a multistep chemical synthesis that is often low yielding, requires costly reagents, and produce diastereoisomeric mixtures. Synthetic biology is an alternative approach to assemble biosynthetic pathways in heterologous host strains using DNA engineering tools. Recently, thaxtomin derivatives, which are diketopiperazine phytotoxins that inhibit cellulose biosynthesis, were successfully biosynthesized using this approach. Herein, genes encoding for thaxtomin nonribosomal peptide synthetases (NRPS) from *Streptomyces scabies,* and promiscuous tryptophan synthase (TrpS) from *Salmonella typhimurium* were assembled as novel artificial pathway in host *Streptomyces albus*, as shown in Fig. 4. Consequently, the engineered strain was fed by indole derivatives to synthesize tryptophan intermediates. In a single fermentation step, these are in turn incorporated by the NRPS to generate various thaxtomins with different functionalities instead of the nitro group (Winn et al. 2018).

Pesticides: insecticides, acaricides, and nematocidal agents

Other than weeds and phytopathogenic microbes, pests such as insects, mites, and soil nematodes, also contribute to harmful actions that reduce crop yield and quality. These can occur during plant growth, post-harvest storage, or transportation. For

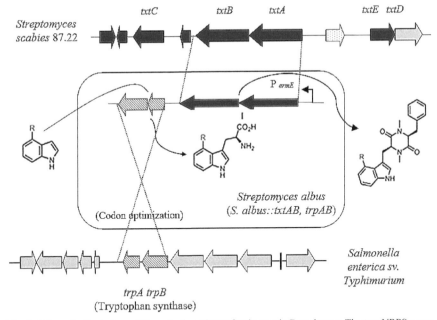

Fig. 4. Schematic illustration of biosynthetic pathway for thaxtomin D analogues. The two NRPS genes from *Streptomyces scabies* were optimized into a synthetic operon that has codon optimized TrpS genes from *Salmonella typhimurium*, which was heterologously expressed in *Streptomyces albus*. Direct feeding of indole analogues to this strain resulted in thaxtomin D derivatives being produced *in vivo*.

this reason, there is an urgent need for excellent and safe pesticides to increase food production and prevent crop loss in the near future. Although the field of plant biotechnology is rapidly developing, it will take a decade or more for genetically engineered insect-tolerant crops to be bred and successfully planted and produced in widespread areas or systematic plant factories. Alternatively, in order to search for less toxic and biodegradable alternatives to chemical pest-control agents, several bioactive agents with pesticidal traits have been studied and characterized from microbes (Djenane et al. 2017).

Bioactive Compounds 1 and 2, isolated from *Streptomyces avermitilis* NEAU1069, showed acaricidal activity against adult mites with a mortality rate of > 90%, and against mite eggs with unhatched egg rates of < 60% at concentrations of 30 and 100 microg/mL, respectively. However, only compound 2 (a hydroxylated derivative at C-23 of 1) exhibited a nematocidal activity against *Caenorhabditis elegans* with an immobility rate of > 90%, at 10 microg/mL concentration (Wang et al. 2010). On the other hand, metabolites in ethyl acetate extract of soil-derived *Streptomyces hydrogenans* DH16, demonstrated anti-insecticidal potential against *Spodoptera litura* at 1600 μg/ml as the highest concentration. The treatment resulted in 70% larval, 66.66% prepupal, and 100% pupal mortality, along with morphological abnormalities and decline in adult emergence, adult longevity, and % hatching. The LC50 and LC90 values were found to be 1337.384 and 2070.516 μg/ml, respectively. Moreover, it is important to mention that several metabolites such as avermectin, milbemycin, spinosad and prasinons have been registered as potential, ecofriendly pesticidal agents (Kaur et al. 2014). Other natural and active insecticides include spinosad, isolated from soil actinomycete *Saccharopolyspora spinosa,* and abamectin, isolated from *Streptomyces avermitilis.* Both compounds are neurotoxins that affect α-amino butyric acid (GABA) receptors. Abamectin also has acaricidal and nematocidal activities (Mossa et al. 2018).

Castaneda-Alvarez and her researchers reported for the first time the nematicidal effects of exoenzymes and metabolites from rizosphere-associated *Bacillus amyloliquefaciens* FR203A on *Xiphinema index.* This strain possessed strong chitinase and lipolytic activities, and exhibited 100% mortality rate on *X. index* juveniles within 72 h of exposure (Castaneda-Alvarez et al. 2016). Soil-derived *Enterobacter asburiae* HK169 also displayed potent nematicidal activity against *Meloidogyne incognita,* the root-knot nematode. The cell-free culture filtrate killed all *M. incognita* juveniles within 48 h. Although proteolytic enzymes might have contributed to the nematicidal activity, gene clusters analysis revealed gene loci for secondary metabolites such as aerobactin, enterobactin, and aryl polyene (Oh et al. 2018). Other *Bacillus* species such as *Bacillus thuringiensis,* isolated from five ecological niches (sediment, rhizospheric and non-rhizospheric soils, grain storage, and dead insects) in Algeria, can be considered as potential candidates for insecticidal biosources. The molecular screening revealed that 15 strains of sampled *B. thuringiensis* possessed all cry genes (*cry1, cry2,* and *cry9*) along with exochitinase, endochitinase and *vip3* genes. These gene families code for lepidopteran-active toxins and entomopathogenic proteins, which are currently investigated for entomocidal control (Djenane et al. 2017). Furthermore, soil *Bacillus* sp. strain FCC41, identified in *Bacillus wiedmannii* cluster, was also named as *B. wiedmannii* biovar *thuringiensis* due to the presence of *cry*-like genes (*cry24Ca, cry4, cry52,* and *cry41* l) encoded in plasmids of different sizes. These encode for cry proteins which known for mosquitocidal activities, which was exhibited against *Aedes aegypti, Aedes (Ochlerotatus) albifasciatus, Culex apicinus, Culex pipiens,* and *Culex quinquefasciatus* (Lazarte et al. 2018).

Interestingly, pesticidal bioactive compounds from fungi were also investigated. Endophytic fungus *Acremonium vitellinum* isolated from marine alga yielded three chloramphenicol derivatives (compounds **1-3**), which showed insecticidal activities against *Helicoverpa armigera,* the cotton bollworm. Compound **2** exhibited significant effects with an LC_{50} value of 0.56 ± 0.03 mg/mL, as compared to weak activities by compounds **1** and **3** with LC_{50} values of 0.93 ± 0.05 and 0.91 ± 0.06 mg/mL, respectively. The positive control, matrine, had an LC_{50} value of 0.24 ± 0.01 mg/mL. Moreover, compound **2** had inhibitory effects on important enzymes in *H. armigera* such as glutathione S-transferase (GST), catalase (CAT), and acetylcholinesterase (AChE). This suggests the molecular mechanism of action accountable for this insecticidal activity (Chen et al. 2018c). Pathogenic soil fungus *Lecanicillium muscarium* was shown to have wide-range insecticidal activities such as against larvae of *Spodoptera litura* (77% mortalities at $\approx 10^8$ conidia/mL) and nymphs of *Ricania simulans* (74.76% mortalities at 1×10^7 conidia/ml). Studies stated that this entomopathogenicity is caused by infecting spores to the insect, which leads to penetration and inside colonization with consequent host death (Mossa et al. 2018).

Plant growth modulators

Phytohormones are essential signaling molecules playing various roles in the physiological development of plants. For example, Gibberellins (GAs) stimulate cell division and elongation, and their exogenous application enhances plant growth under stress conditions such as drought, high salinity, and extreme temperatures. On the other hand, indole-3-acetic acid (IAA) also has a crucial function in cell elongation, division, and enlargement. Endophtic bacteria and fungi have been proven to secrete plant growth modulators as bioactive agents to assist plants in resisting biotic and abiotic stresses, which could lead to devastating agricultural losses. Another important property of microbial endophytes is organic acids production. These recycle minerals from soil to plants and deactivate metal ions to stable chelating complexes, thus regulating plant physiological processes. In addition, organic acids are indispensable in weathering and biodeterioration (Waqas et al. 2014, Sathya et al. 2017). Other potentially vital agrochemical metabolites isolated from microbes along with plant growth modulators are summarized in Table 3.

Streptomyces sp. CB-75 isolated from Chinese diseased soils in Hainan, which was characterized as *Streptomyces spectabilis*, possessed (*Z*)-13-docosenamide as a major constituent in its crude extract. This strain demonstrated a potent growth-promoting activity on banana plants as chlorophyll content was enhanced by 88.24%, leaf area by 88.24%, root length by 90.49%, root diameter by 136.17%, plant height by 61.78%, and stem growth by 50.98%. In addition, fresh shoot and root weights were improved by 195.33 and 113.33%, and dry shoot and root weights were improved by 195.33, and 113.33%, respectively, when compared to control treatment with fermentation broth (Chen et al. 2018b).

Volatile organic compounds (VOCs) isolated from *Streptomyces* species, abundant in disease-suppressive soils to *Rhizoctonia solani,* significantly enhanced plant root and shoot biomass in *Arabidopsis thaliana* seedlings. Among tested isolates, *Streptomyces* strains W47 and W62 along with *Streptomyces lividans* 1326 have resulted in the largest increase of plant biomass (Cordovez et al. 2015). Similar findings were achieved from a study on *Bacillus mojavensis* RRC101, when 1/2x MS medium was used for bacterial growth. According to GC–MS analysis coupled with solid phase microextraction (SPME), the highly detected growth modulating compounds were identified as 3-hydroxy-2-butanone (acetoin) and 2,3-butanediol. These VOCs are produced by other species such as *Bacillus subtilis* and *Bacillus amyloliquefaciens* (Rath et al. 2018). Additionally, culture filtrate of *Enterobacter asburiae* HK169 reduced gall formation by 66% in tomatoes, whereas it increased shoot and root weights by 160% and 251%, respectively, compared with negative control. Biosynthetic genes for two siderophores aerobactin and enterobactin, and aryl polyene were sequenced, which suggest the reason behind this growth promoting activity (Oh et al. 2018).

Table 3. Summary of recently isolated bioactive compounds and their agrochemical activities.

Bioactive compound	Microorganism/producer	Source	Biological activities	Reference
1-acetyl-β-carboline Bacillomycin and Fengycin	*Bacillus* species	Tunisian arid areas	Fungicide	(Nasfi et al. 2018)
NRPS and PKS-I, (Z)-13-docosenamide	*Streptomyces* sp. CB-75	Banana rhizosphere soil from China	Fungicide, plant growth promoting activity	(Chen et al. 2018b)
Streptotricins D, E and F	*Streptomyces* sp. A20	Rhizospheric soils from Tolima, Colombia	Bacteriocide	(Suárez-Moreno et al. 2019)
Germination Arrest Factor (GAF)	*Pseudomonas fluorescens* strain WH6	Rhizosphere soil	Bacteriocide, Herbicide	(Harding and Raizada 2015)
Macrocidins, Chenopodolin	*Phoma macrostoma, Phoma chenopodicola*	Field isolates	Herbicides	(Harding and Raizada 2015)
Penicillic acid	*Penicillium canescens* and *Penicillium cyclopium*		Herbicide	(Bazioli et al. 2017)
Phomentrioloxin	*Phomopsis* sp.	Fungal pathogen for *Carthamus lanatus*	Herbicides	(Massi et al. 2018)
Cytochalasans and nonenolides	*Ascochyta, Phoma,* and *Stagonospora* spp.			
Papyracillic acid and agropyrenol	*Ascochyta agropyrina* var. *nana*	Fungal pathogen of *Elytrigia repens*		
Chenopodolin	*Phoma chenopodiicola*	Fungal pathogen for *Chenopodium album*		
Cochliotoxin, radicinin, and 3-epi-radicinin	*Cochliobolus australiensis*	Buffelgrass		
Sphaeropsidone, Sphaeropsidin A	*Diplodia cupressi*	Diseased cypress in the Mediterranean area	Herbicide, Bacteriocide	(Massi et al. 2018)
Lipopolysaccharide and extracellular peptides	*Pseudomonas fluorescens* strain D7	Rhizospheres of *Triticum aestivum*	Herbicides	(Harding and Raizada 2015)

Compound	Organism	Source	Activity	Reference
Spinosad, abamectin	*Saccharopolyspora spinosa, Streptomyces avermitilis*	Soil	Insecticide, acaricidal and nematocidal activities	(Mossa et al. 2018)
Aerobactin, enterobactin, and aryl polyene	*Enterobacter asburiae* HK169	Soil	Nematicidal activity, Plant growth promoting activity (Increased shoot and root weights) Reduced gall formation	(Oh et al. 2018)
Chloramphenicol derivatives	*Acremonium vitellinum*	Marine algae	Insecticidal activities	(Chen et al. 2018c)
Volatile organic compounds (VOCs)	*Streptomyces* spp. W47, W62 and *Streptomyces lividans* 1326	Disease-suppressive soils to *Rhizoctonia solani*	Plant growth promoting activity (Increased biomass)	(Cordovez et al. 2015)
3-hydroxy-2-butanone (acetoin) and 2,3-butanediol	*Bacillus mojavensis* RRC101	Rhizosphere soil	Plant growth modulators	(Rath et al. 2018)
Pteridic acids A and B	*Streptomyces hygroscopicus* TP-A045	Stem of bracken, *Pteridium aquilinum*, from Japan	Auxin-like activity and root elongation induction	(Sathya et al. 2017)
ACC (1-aminocyclopropane-1-carboxylate) deaminase, IAA	*Streptomyces filipinensis*	Tomato rhizosphere soil	Increased growth promotion	(Vurukonda et al. 2018)
Gibberellins (GA$_1$, GA$_3$, GA$_4$, GA$_7$, and GA$_9$), IAA Oxalic, quinic, malic, and citric acids	*Aspergillus fumigates, Chrysosporium pseudomerdarium, Phoma glomerata,* and *Paecilomyces formosus*	*Cucumis sativus* and *Glycine max* (L.) Merr. fungal endophytes	Promoted growth of shoot length, chlorophyll contents, and biomass	(Waqas et al. 2014)

It was also reported that IAA producing *Streptomyces olivaceoviridis,* *Streptomyces atrovirens, Streptomyces rimosus, Streptomyces rochei,* and *Streptomyces viridis* showed enhanced seed germination and root growth, whereas IAA from *Streptomyces atroolivaceus* induced cell differentiation, sporulation, and hyphal elongation. Moreover, pteridic acids A and B produced by *Streptomyces hygroscopicus* TP-A045 exhibited auxin-like activity and induced root elongation in kidney beans. Yet, the highest IAA producing actinobacteria were identified as *Nocardiopsis* sp., yielding 222.75 ppm (Sathya et al. 2017). In a detailed review by Vurukonda et al. (2018), other IAA producing actinomycete included *Streptomyces filipinensis,* which also produced ACC (1-aminocyclopropane-1-carboxylate) deaminase, that has led to increased growth promotion, when compared to *Streptomyces atrovirens* that produced only ACC deaminase. Therefore, producing multiple plant hormones by a single *Streptomycetes* was shown to be more effective. Besides, production and the excretion of enterobactin, an iron-chelating compound, by *Streptomyces* species from Western Australian soils was also reported (Vurukonda et al. 2018).

Interestingly, *Cucumis sativus* and *Glycine max* (L.) Merr. fungal endophytes promoted growth of shoot length, chlorophyll contents, and biomass of both wild-type and mutant rice (*Oryza sativa*). These included *Aspergillus fumigates, Chrysosporium pseudomerdarium, Phoma glomerata,* and *Paecilomyces formosus,* in addition to *Paecilomyces, Penicillium* species. According to chromatographic analysis of culture filtrates, bioactive chemical constituents were identified as gibberellins (GA_1, GA_3, GA_4, GA_7, and GA_9), IAA, as well as oxalic, quinic, malic, and citric acids (Waqas et al. 2014).

Conclusions

Bioactive compounds that are naturally produced by various species of bacteria and fungi contribute to a wide range of significant biological activities. These include anti-microbial, anti-cancer, anti-inflammatory, anti-oxidant, enzyme inhibitory, and immunomodulatory activities, in addition to agrochemicals such as herbicides, pesticides, and plant growth modulators. Recent research has also pointed out the feasibility of these isolated compounds compared to currently used, synthesized chemicals. Among the advantages of microbial secondary metabolites are their environmentally friendly nature and lower toxicity towards other livings things including humans. Therefore, biotechnological research must continue to investigate these compounds and exploit them to provide better therapies and enhance agriculture in near future.

Acknowledgments

Author wishes to thank her family and professors from Qatar University for their encouragements to write this chapter contributing to the book.

References

Abdel-Shakour E. H., Qari, S. H. M., Beltagy, E. A., Abou El-Ela, G. M., Fahmy, N. M. and Reffat, B. M. 2015. Assessment of antimicrobial, antioxidant and enzyme inhibitory activities of *Streptomyces* sp. Nyr04 (KT074931) isolated from mangrove sediment in Red Sea (Egypt). Wulfenia 22(10): 472–481.

Ali, L., Khan, A. L., Hussain, J., Al-Harrasi, A., Waqas, M., Kang, S. M. et al. 2016. Sorokiniol: a new enzymes inhibitory metabolite from fungal endophyte *Bipolaris sorokiniana* LK12. BMC Microbiology 16: 103. doi:10.1186/s12866-016-0722-7.

An, C. L., Kong, F. D., Ma, Q. Y., Xie, Q. Y., Yuan, J. Z., Zhou, L. M. et al. 2018. Chemical constituents of the marine-derived fungus *Aspergillus* sp. SCS-KFD66. Marine Drugs 16(12): 468. doi:10.3390/md16120468.

Andreolli, M., Zapparoli, G., Angelini, E., Lucchetta, G., Lampis, S. and Vallini, G. 2019. *Pseudomonas protegens* MP12: A plant growth-promoting endophytic bacterium with broad-spectrum antifungal activity against grapevine phytopathogens. Microbiological Research 219: 123–131. doi:10.1016/j.micres.2018.11.003.

Awad, N. E., Kassem, H. A., Hamed, M. A., El-Feky, A. M., Elnaggar, M., Mahmoud, K. et al. 2018. Isolation and characterization of the bioactive metabolites from the soil derived fungus *Trichoderma viride*. Mycology 9(1): 70–80. doi:10.1080/21501203.2017.1423126.

Balasubramanian, S., Othman, E. M., Kampik, D., Stopper, H., Hentschel, U., Ziebuhr, W. et al. 2017. Marine sponge-derived *Streptomyces* sp. SBT343 extract inhibits Staphylococcal biofilm formation. Frontiers in Microbiology 8: 236. doi:10.3389/fmicb.2017.00236.

Bazioli, J. M., Amaral, L., Fill, T. P. and Rodrigues-Filho, E. 2017. Insights into *Penicillium brasilianum* secondary metabolism and its biotechnological potential. Molecules (Basel, Switzerland) 22(6): 858. doi:10.3390/molecules22060858.

Ben Mefteh, F., Daoud, A., Chenari Bouket, A., Thissera, B., Kadri, Y., Cherif-Silini, H. et al. 2018. Date palm trees root-derived endophytes as fungal cell factories for diverse bioactive metabolites. International Journal of Molecular Sciences 19(7): 1986. doi:10.3390/ijms19071986.

Bennur, T., Ravi Kumar, A., Zinjarde, S. and Javdekar, V. 2016. *Nocardiopsis* species: a potential source of bioactive compounds. Journal of Applied Microbiology 120: 1–16. doi:10.1111/jam.12950.

Bérdy, J. 2005. Bioactive microbial metabolites. The Journal of Antibiotics 58(1): 1–26. Doi: 10.1038/ja.2005.1.

Braña, A. F., Sarmiento-Vizcaíno, A., Osset, M., Pérez-Victoria, I., Martín, J., de Pedro, N. et al. 2017. Lobophorin K, a new natural product with cytotoxic activity produced by *Streptomyces* sp. M-207 associated with the deep-sea coral Lophelia pertusa. Marine Drugs 15(5): 144. doi:10.3390/md15050144.

Buedenbender, L., Robertson, L. P., Lucantoni, L., Avery, V. M., Kurtböke, D. İ. and Carroll, A. R. 2018. HSQC-TOCSY fingerprinting-directed discovery of antiplasmodial polyketides from the marine ascidian-derived *Streptomyces* sp. (USC-16018). Marine Drugs 16(6): 189. doi:10.3390/md16060189.

Campos, F. F., Ramos, J. P., Oliveira, D. M., Alves, T. M., Souza-Fagundes, E. M., Zani, C. L. et al. 2017. *In vitro* leishmanicidal, antibacterial and antitumour potential of anhydrocochlioquinone A obtained from the fungus *Cochliobolus* sp. Journal of Biosciences 42(4): 657–664. doi:10.1007/s12038-017-9718-1.

Castaneda-Alvarez, C., Prodan, S., Rosales, I. and Aballay, E. 2016. Exoenzymes and metabolites related to the nematicidal effect of rhizobacteria on *Xiphinema index* Thorne & Allen. Journal of Applied Microbiology 120: 413–424. doi:10.1111/jam.12987.

Charousová, I., Medo, J., Hleba, L. and Javoreková, S. 2018. *Streptomyces globosus* DK15 and *Streptomyces ederensis* ST13 as new producers of factumycin and tetrangomycin antibiotics. Brazilian Journal of Microbiology 49(4): 816–822.

Chen, M., Chai, W., Song, T., Ma, M., Lian, X. Y. and Zhang, Z. 2018a. Anti-glioma natural products downregulating tumor glycolytic enzymes from marine actinomycete *Streptomyces* sp. ZZ406. Scientific Reports 8(1): 72. doi:10.1038/s41598-017-18484-7.

Chen, Y., Zhou, D., Qi, D., Gao, Z., Xie, J. and Luo, Y. 2018b. Growth promotion and disease suppression ability of a *Streptomyces* sp. CB-75 from banana rhizosphere soil. Frontiers in Microbiology 8: 2704. doi:10.3389/fmicb.2017.02704.

Chen, D., Zhang, P., Liu, T., Wang, X. F., Li, Z. X., Li, W. et al. 2018c. Insecticidal activities of chloramphenicol derivatives isolated from a marine alga-derived endophytic fungus, *Acremonium vitellinum*, against the Cotton Bollworm, *Helicoverpa armigera* (Hübner) (Lepidoptera: Noctuidae). Molecules (Basel, Switzerland). 23(11): 2995. doi:10.3390/molecules23112995.

Cheng, Z., Xu, W., Liu, L., Li, S., Yuan, W., Luo, Z. et al. 2018. Peniginsengins B⁻ E, New Farnesylcyclohexenones from the deep sea-derived fungus *Penicillium* sp. YPGA11. Marine Drugs 16(10): 358. doi:10.3390/md16100358.

Córdova-Dávalos, L., Escobedo-Chávez, K. and Evangelista-Martínez, Z. 2018. Inhibition of *Candida albicans* cell growth and biofilm formation by a bioactive extract produced by soil *Streptomyces* strain GCAL-25. Archives of Biological Sciences Arhiv Za Bioloske Nauke 70(2): 387–396. doi:10.2298/abs170908057c.

Cordovez, V., Carrion, V. J., Etalo, D. W., Mumm, R., Zhu, H., van Wezel, G. P. et al. 2015. Diversity and functions of volatile organic compounds produced by *Streptomyces* from a disease-suppressive soil. Frontiers in Microbiology 6: 1081. doi:10.3389/fmicb.2015.01081.

Cot, M., Ray, A., Gilleron, M., Vercellone, A., Larrouy-Maumus, G., Armau, E. et al. 2011. Lipoteichoic acid in *Streptomyces hygroscopicus*: structural model and immunomodulatory activities. PloS One 6(10): e26316. doi:10.1371/journal.pone.0026316.

Cota, B. B., Tunes, L. G., Maia, D., Ramos, J. P., Oliveira, D. M., Kohlhoff, M. et al. 2018. Leishmanicidal compounds of *Nectria pseudotrichia*, an endophytic fungus isolated from the plant *Caesalpinia echinata* (Brazilwood). Memorias do Instituto Oswaldo Cruz 113(2): 102–110.

Debbab, A., Aly, A. H., Lin, W. H. and Proksch, P. 2010. Bioactive compounds from marine bacteria and fungi. Microbial Biotechnology 3(5): 544–563. doi:10.1111/j.1751-7915.2010.00179.x.

Demers, D. H., Knestrick, M. A., Fleeman, R., Tawfik, R., Azhari, A., Souza, A. et al. 2018. Exploitation of mangrove endophytic fungi for infectious disease drug discovery. Marine Drugs 16(10): 376. doi:10.3390/md16100376.

Dholakiya, R. N., Kumar, R., Mishra, A., Mody, K. H. and Jha, B. 2017. Antibacterial and antioxidant activities of novel *Actinobacteria* strain isolated from gulf of Khambhat, Gujarat. Frontiers in Microbiology 8: 2420. doi:10.3389/fmicb.2017.02420.

Djenane, Z., Nateche, F., Amziane, M., Gomis-Cebolla, J., El-Aichar, F., Khorf, H. et al. 2017. Assessment of the antimicrobial activity and the entomocidal potential of *Bacillus thuringiensis* isolates from algeria. Toxins 9(4): 139. doi:10.3390/toxins9040139.

El-Hadedy, D. E., Mostafa, E. E. and Saad, M. M. 2015. *Streptomyces lavendulae* protease inhibitor: purification, gene overexpression, and 3-dimensional structure. Journal of Chemistry 2015: Article ID 963041, 1–9. doi:10.1155/2015/963041.

El Sayed, O. H., Asker, M., Swelim, M. A., Abbas, I. H., Attwa, A. I. and El Awady, M. E. 2016. Production of hydroxy marilone C as a bioactive compound from *Streptomyces badius*. Journal Genetic Engineering & Biotechnology 14(1): 161–168.

Górska, S., Hermanova, P., Ciekot, J., Schwarzer, M., Srutkova, D., Brzozowska, E. et al. 2016. Chemical characterization and immunomodulatory properties of polysaccharides isolated from probiotic *Lactobacillus casei* LOCK 0919. Glycobiology 26(9): 1014–1024. doi:10.1093/glycob/cww047.

Ha, T. M., Ko, W., Lee, S. J., Kim, Y. C., Son, J. Y., Sohn, J. H. et al. 2017. Anti-inflammatory effects of curvularin-type metabolites from a marine-derived fungal strain *Penicillium* sp. SF-5859 in lipopolysaccharide-induced RAW264.7 macrophages. Marine Drugs 15(9): 282. doi:10.3390/md15090282.

Hamzah, T., Lee, S. Y., Hidayat, A., Terhem, R., Faridah-Hanum, I. and Mohamed, R. 2018. Diversity and characterization of endophytic fungi isolated from the tropical mangrove species, *Rhizophora mucronata*, and identification of potential antagonists against the soil-borne fungus, *Fusarium solani*. Frontiers in Microbiology 9: 1707. doi:10.3389/fmicb.2018.01707.

Harding, D. P. and Raizada, M. N. 2015. Controlling weeds with fungi, bacteria and viruses: a review. Frontiers in Plant Science 6: 659. doi:10.3389/fpls.2015.00659.

Janardhan, A., Kumar, A. P., Viswanath, B., Saigopal, D. V. and Narasimha, G. 2014. Production of bioactive compounds by actinomycetes and their antioxidant properties. Biotechnology Research International 2014: 217030. doi:10.1155/2014/217030.

Kang, H. H., Zhang, H. B., Zhong, M. J., Ma, L. Y., Liu, D. S., Liu, W. Z. et al. 2018. Potential antiviral xanthones from a coastal saline soil fungus *Aspergillus iizukae*. Marine Drugs 16(11): 449. doi:10.3390/md16110449.

Kaur, T., Vasudev, A., Sohal, S. K. and Manhas, R. K. 2014. Insecticidal and growth inhibitory potential of *Streptomyces hydrogenans* DH16 on major pest of India, *Spodoptera litura* (Fab.) (*Lepidoptera: Noctuidae*). BMC Microbiology 14: 227. doi:10.1186/s12866-014-0227-1.

Kodani, S., Komaki, H., Hemmi, H., Miyake, Y., Kaweewan, I. and Dohra, H. 2018. Streptopeptolin, a cyanopeptolin-type peptide from *Streptomyces olivochromogenes*. ACS Omega 3(7): 8104–8110. doi:10.1021/acsomega.8b01042.

Kumar, S. and Rao, K. 2018. Efficacy of alpha glucosidase inhibitor from marine actinobacterium in the control of postprandial hyperglycaemia in streptozotocin (STZ) induced diabetic male Albino Wister Rats. Iranian Journal of Pharmaceutical Research: IJPR 17(1): 202–214.

Kumari, M., Taritla, S., Sharma, A. and Jayabaskaran, C. 2018. Antiproliferative and antioxidative bioactive compounds in extracts of marine-derived endophytic fungus *Talaromyces purpureogenus*. Frontiers in Microbiology 9: 1777. doi:10.3389/fmicb.2018.01777.

Lalitha, C., Raman, T., Rathore, S. S., Ramar, M., Munusamy, A. and Ramakrishnan, J. 2017. ASK2 bioactive compound inhibits MDR *Klebsiella pneumoniae* by antibiofilm activity, modulating macrophage cytokines and opsonophagocytosis. Frontiers in Cellular and Infection Microbiology 7: 346. doi:10.3389/fcimb.2017.00346.

Lazarte, J. N., Lopez, R. P., Ghiringhelli, P. D. and Berón, C. M. 2018. *Bacillus wiedmannii* biovar *thuringiensis*: A specialized mosquitocidal pathogen with plasmids from diverse origins. Genome Biology and Evolution 10(10): 2823–2833. doi:10.1093/gbe/evy211.

Li, H., Huang, H., Hou, L., Ju, J. and Li, W. 2017. Discovery of antimycin-type depsipeptides from a *wbl* Gene mutant strain of deepsea-derived *Streptomyces somaliensis* SCSIO ZH66 and their effects on pro-inflammatory cytokine production. Frontiers in Microbiology 8: 678. doi:10.3389/fmicb.2017.00678.

Li, F., Chen, D., Lu, S., Yang, G., Zhang, X., Chen, Z. et al. 2018. Anti-influenza a viral butenolide from *Streptomyces* sp. Smu03 inhabiting the intestine of *Elephas maximus*. Viruses 10(7): 356. doi:10.3390/v10070356.

Liang, T. W., Chen, W. T., Lin, Z. H., Kuo, Y. H., Nguyen, A. D., Pan, P. S. et al. 2016. An amphiprotic novel chitosanase from *Bacillus mycoides* and its application in the production of chitooligomers with their antioxidant and anti-inflammatory evaluation. International Journal of Molecular Sciences 17(8): 1302. doi:10.3390/ijms17081302.

Liu, Y., Chen, S., Liu, Z., Lu, Y., Xia, G., Liu, H. et al. 2015. Bioactive metabolites from mangrove endophytic fungus *Aspergillus* sp. 16-5B. Marine Drugs 13(5): 3091–3102. doi:10.3390/md13053091.

Ma, J., Cao, B., Liu, C., Guan, P., Mu, Y., Jiang, Y. et al. 2018. Actinofuranones D-I from a lichen-associated actinomycetes, *Streptomyces gramineus*, and their anti-inflammatory effects. Molecules (Basel, Switzerland) 23(9): 2393. doi:10.3390/molecules23092393.

Mahmoudi, F., Baradaran, B., Dehnad, A., Shanehbandi, D., Khosroshahi, L. D. and Aghapour, M. 2016. The immunomodulatory activity of secondary metabolites isolated from *Streptomyces calvus* on human peripheral blood mononuclear cells. British Journal of Biomedical Science 73(3): 97–103. Doi: 10.1080/09674845.2016.1188476.

Managamuri, U., Vijayalakshmi, M., Ganduri, V., Rajulapati, S. B., Bonigala, B., Kalyani, B. S. et al. 2017. Isolation, identification, optimization, and metabolite profiling of *Streptomyces sparsus* VSM-30. 3 Biotech. 7(3): 217. Doi:10.1007/s13205-017-0835-1.

Masi, M., Nocera, P., Reveglia, P., Cimmino, A. and Evidente, A. 2018. Fungal metabolite antagonists of plant pests and human pathogens: Structure-activity relationship studies. Molecules (Basel, Switzerland). 23(4): 834. Doi:10.3390/molecules23040834.

Missoum, A. and Al-Thani, R. 2017. Production of antimicrobial agents by *Bacillus* spp. isolated from Al-Khor coast soils, Qatar. African Journal of Microbiology Research 11(41): 1510–1519.

Mossa, A. H., Mohafrash, S. and Chandrasekaran, N. 2018. Safety of natural insecticides: Toxic effects on experimental animals. BioMed. Research International 2018: 4308054. doi:10.1155/2018/4308054.

Nasfi, Z., Busch, H., Kehraus, S., Linares-Otoya, L., König, G. M., Schäberle, T. F. et al. 2018. Soil bacteria isolated from tunisian arid areas show promising antimicrobial activities against gram-negatives. Frontiers in Microbiology 9: 2742. doi:10.3389/fmicb.2018.02742.

Oh, M., Han, J. W., Lee, C., Choi, G. J. and Kim, H. 2018. Nematicidal and plant growth-promoting activity of *Enterobacter asburiae* HK169: Genome analysis provides insight into its biological activities. Journal of Microbiology and Biotechnology 28(6): 968–975. doi:10.4014/jmb.1801.01021.

Patel, P., Shah, R., Joshi, B., Ramar, K. and Natarajan, A. 2019. Molecular identification and biocontrol activity of sugarcane rhizosphere bacteria against red rot pathogen *Colletotrichum falcatum*. Biotechnology Reports (Amsterdam, Netherlands). 21: e00317. doi:10.1016/j.btre.2019.e00317.

Raekiansyah, M., Mori, M., Nonaka, K., Agoh, M., Shiomi, K., Matsumoto, A. et al. 2017. Identification of novel antiviral of fungus-derived brefeldin A against dengue viruses. Tropical Medicine and Health. 45: 32. doi:10.1186/s41182-017-0072-7.

Raghava Rao, K. V., Mani, P., Satyanarayana, B. and Raghava Rao, T. 2017. Purification and structural elucidation of three bioactive compounds isolated from *Streptomyces coelicoflavus* BC 01 and their biological activity. 3 Biotech. 7(1): 24. doi:10.1007/s13205-016-0581-9.

Rath, M., Mitchell, T. and Gold, S. 2018. Volatiles produced by *Bacillus mojavensis* RRC101 act as plant growth modulators and are strongly culture-dependent. Microbiological Research. 208: 76–84. doi:10.1016/j.micres.2017.12.014.

Rodríguez, V., Martín, J., Sarmiento-Vizcaíno, A., de la Cruz, M., García, L. A., Blanco, G. et al. 2018. Anthracimycin B, a potent antibiotic against gram-positive bacteria isolated from cultures of the deep-sea actinomycete *Streptomyces cyaneofuscatus* M-169. Marine Drugs 16(11): 406. doi:10.3390/md16110406.

Rogozhin, E. A., Sadykova, V. S., Baranova, A. A., Vasilchenko, A. S., Lushpa, V. A., Mineev, K. S. et al. 2018. A novel lipopeptaibol emericellipsin a with antimicrobial and antitumor activity produced by the extremophilic fungus *Emericellopsis alkalina*. Molecules (Basel, Switzerland). 23(11): 2785. doi:10.3390/molecules23112785.

Sathya, A., Vijayabharathi, R. and Gopalakrishnan, S. 2017. Plant growth-promoting actinobacteria: a new strategy for enhancing sustainable production and protection of grain legumes. 3 Biotech. 7(2): 102. doi:10.1007/s13205-017-0736-3.

Selama, O., Amos, G. C., Djenane, Z., Borsetto, C., Laidi, R. F., Porter, D. et al. 2014. Screening for genes coding for putative antitumor compounds, antimicrobial and enzymatic activities from haloalkalitolerant and haloalkaliphilic bacteria strains of Algerian Sahara soils. BioMed. Research International 2014: 317524. doi:10.1155/2014/317524.

Siddharth, S. and Vittal, R. R. 2018. Evaluation of antimicrobial, enzyme inhibitory, antioxidant and cytotoxic activities of partially purified volatile metabolites of marine *Streptomyces* sp. S2A. Microorganisms 6(3): 72. doi:10.3390/microorganisms6030072.

Shah, M., Deshmukh, S. K., Verekar, S. A., Gohil, A., Kate, A. S., Rekha, V. et al. 2015. Anti-inflammatory properties of mutolide isolated from the fungus *Lepidosphaeria* species (PM0651419). SpringerPlus 4: 706. doi:10.1186/s40064-015-1493-6.

Shahzad, R., Latif Khan, A., Ali, L., Bilal, S., Imran, M., Choi, K. S. et al. 2018. Characterization of new bioactive enzyme inhibitors from endophytic *Bacillus amyloliquefaciens* RWL-1. Molecules (Basel, Switzerland) 23(1): 114. doi:10.3390/molecules23010114.

Sharma, G., Dang, S., Gupta, S. and Gabrani, R. 2018. Antibacterial activity, cytotoxicity, and the mechanism of action of bacteriocin from *Bacillus subtilis* GAS101. Medical Principles and Practice: International Journal of the Kuwait University, Health Science Centre 27(2): 186–192.

Sheir, D. H. and Hafez, M. A. 2017. Antibiofilm activity of *Streptomyces toxytricini* Fz94 against *Candida albicans* ATCC 10231. Microbial Biosystems 2(1): 26–39. doi:10.21608/mb.2017.5255.

Shim, D. W., Shin, H. J., Han, J. W., Shin, W. Y., Sun, X., Shim, E. J. et al. 2015. Anti-inflammatory effect of Streptochlorin via TRIF-dependent signaling pathways in cellular and mouse models. International Journal of Molecular Sciences 16(4): 6902–6910. doi:10.3390/ijms16046902.

Singh, V., Haque, S., Khare, S., Tiwari, A. K., Katiyar, D., Banerjee, B. et al. 2018. Isolation and purification of antibacterial compound from *Streptomyces levis* collected from soil sample of north India. PloS One 13(7): e0200500. doi:10.1371/journal.pone.0200500.

So, K. K., Chun, J. and Kim, D. H. 2018. Antimicrobial and antitumor photodynamic effects of phleichrome from the phytopathogenic fungus *Cladosporium Phlei*. Mycobiology 46(4): 448–451. doi:10.1080/12298093.2018.1551599.

Srivastava, V. and Dubey, A. K. 2016. Anti-biofilm activity of the metabolites of *Streptomyces chrestomyceticus* strain ADP4 against *Candida albicans*. Journal of Bioscience and Bioengineering 122(4): 434–440. doi:10.1016/j.jbiosc.2016.03.013.

Starosila, D., Rybalko, S., Varbanetz, L., Ivanskaya, N. and Sorokulova, I. 2017. Anti-influenza activity of a *Bacillus subtilis* probiotic strain. Antimicrobial Agents and Chemotherapy 61(7): e00539–17. doi:10.1128/AAC.00539-17.

Suárez-Moreno, Z. R., Vinchira-Villarraga, D. M., Vergara-Morales, D. I., Castellanos, L., Ramos, F. A., Guarnaccia, C. et al. 2019. Plant-growth promotion and biocontrol properties of three *Streptomyces* spp. isolates to control bacterial rice pathogens. Frontiers in Microbiology 10: 290. doi:10.3389/fmicb.2019.00290.

Tan, L. T., Chan, K. G., Chan, C. K., Khan, T. M., Lee, L. H. and Goh, B. H. 2018. Antioxidative potential of a *Streptomyces* sp. MUM292 isolated from mangrove soil. BioMed. Research International 2018: 4823126. doi:10.1155/2018/4823126.

Tan, L. T., Chan, K. G., Pusparajah, P., Yin, W. F., Khan, T. M., Lee, L. H. et al. 2019. Mangrove derived *Streptomyces* sp. MUM265 as a potential source of antioxidant and anticolon-cancer agents. BMC Microbiology 19(1): 38. doi:10.1186/s12866-019-1409-7.

Treasure, U., Eze, P., Umeokoli, B. O., Abbah, C., Okoye, F. and Esimone, C. 2017. Evaluation of antiplasmodial and immunomodulatory activities of extracts of endophytic fungi isolated from four nigerian medicinal plants. PMIO 4. doi: 10.1055/s-0037-1608414.

Utesch, T., de Miguel Catalina, A., Schattenberg, C., Paege, N., Schmieder, P., Krause, E. et al. 2018. A computational modeling approach predicts interaction of the antifungal protein AFP from *Aspergillus giganteus* with fungal membranes via its γ-core motif. mSphere 3(5): e00377–18. doi:10.1128/mSphere.00377-18.

Vurukonda, S., Giovanardi, D. and Stefani, E. 2018. Plant growth promoting and biocontrol activity of *Streptomyces* spp. as endophytes. International Journal of Molecular Sciences 19(4): 952. doi:10.3390/ijms19040952.

Wang, X., Wang, M., Wang, J., Jiang, L., Wang, J. and Xiang, W. 2010. Isolation and identification of novel macrocyclic lactones from *Streptomyces avermitilis* NEAU1069 with acaricidal and nematocidal activity. Journal of Agricultural and Food Chemistry 58(5): 2710–2714. doi:10.1021/jf902496d.

Waqas, M., Khan, A. L. and Lee, I. J. 2014. Bioactive chemical constituents produced by endophytes and effects on rice plant growth. Journal of Plant Interactions 9(1): 478–487. doi: 10.1080/17429145.2013.860562.

Winn, M., Francis, D. and Micklefield, J. 2018. *De novo* biosynthesis of "Non-Natural" thaxtomin phytotoxins. Angewandte Chemie (International ed. in English) 57(23): 6830–6833. doi:10.1002/anie.201801525.

Xie, Y., Li, S., Sun, L., Liu, S., Wang, F., Wen, B. et al. 2018. Fungal immunomodulatory protein from *Nectria haematococca* suppresses growth of human lung adenocarcinoma by inhibiting the PI3K/Akt pathway. International Journal of Molecular Sciences 19(11): 3429. doi:10.3390/ijms19113429.

Yang, B., Sun, W., Wang, J., Lin, S., Li, X. N., Zhu, H. et al. 2018. A new breviane spiroditerpenoid from the marine-derived fungus Penicillium sp. TJ403-1. Marine Drugs 16(4): 110. doi:10.3390/md16040110.

Zhang, S., Zhu, J., Zechel, D. L., Jessen-Trefzer, C., Eastman, R. T., Paululat, T. et al. 2017a. New WS9326A derivatives and one new annimycin derivative with antimalarial activity are produced by *Streptomyces asterosporus* DSM 41452 and its mutant. Chembiochem: A European Journal of Chemical Biology 19(3): 272–279.

Zhang, X., Ye, X., Chai, W., Lian, X. Y. and Zhang, Z. 2016. New metabolites and bioactive actinomycins from marine-derived *Streptomyces* sp. ZZ338. Marine Drugs 14(10): 181. doi:10.3390/md14100181.

Zhang, L., Niaz, S. I., Khan, D., Wang, Z., Zhu, Y., Zhou, H. et al. 2017b. Induction of diverse bioactive secondary metabolites from the mangrove endophytic fungus *Trichoderma* sp. (Strain 307) by co-cultivation with *Acinetobacter johnsonii* (Strain B2). Marine Drugs 15(2): 35. doi:10.3390/md15020035.

Zhang, D., Shu, C., Lian, X. and Zhang, Z. 2018. New antibacterial bagremycins F and G from the marine-derived *Streptomyces* sp. ZZ745. Marine Drugs 16(9): 330. doi:10.3390/md16090330.

Zhou, Y. M., Ju, G. L., Xiao, L., Zhang, X. F. and Du, F. Y. 2018. Cyclodepsipeptides and sesquiterpenes from marine-derived fungus *Trichothecium roseum* and their biological functions. Marine Drugs 16(12): 519. doi:10.3390/md16120519.

Zimmermann, P., Ziesenitz, V. C., Curtis, N. and Ritz, N. 2018. The immunomodulatory effects of macrolides—A systematic review of the underlying mechanisms. Frontiers in Immunology 9: 302. doi:10.3389/fimmu.2018.00302.

CHAPTER 7

Waste Treatments

Aiyoub Shahi,[1,]* *Duygu Nur Arabaci,*[2] *Onat Cinlar*[3] and
Sevcan Aydin[4]

Introduction

Economic and technological development around the world has not only increased the quantity of generated waste but also increased the complexity of produced wastes. Industrial expansions, agricultural activities, rapid urbanization and medical progress have added huge amount of hazardous waste to the nature which potentially pose great risk to the environment and human health (Singh 2013). Biological processes play a main role in the removal of contaminants from the environment. Microbial strains, because of their catabolic veracity, degrade/convert such pollutants (Loss and Yu 2018).

Wastes are undesirable residues with negative value which are not required where they are produced (Singh 2013). Wastes which are produced in different solid, liquid and gaseous forms mostly because of their hazardous properties are needed to be get rid of and discarded (Nelson 2006). Common domestic waste, industrial and commercial waste generated by factories and industrial plants, sewage sludge which is produced as a by-product during industrial and municipal wastewater treatment, agricultural and food processing waste, and mining waste are among the solid waste category (Barbuta et al. 2015). Liquid waste includes fluids such as oils, fats, greases, domestic and industrial wastewater. Gaseous waste originates from various human activities, such as incineration, manufacturing and processing, vehicles and agricultural processes (Singh 2013). In terms of degradability, waste can be divided into two types: biodegradable and non-biodegradable waste. Biodegradable material can be decomposed by bacteria and other organisms while non-biodegradable waste cannot be broken down or degraded by living organisms. Food material, paper waste,

[1] Institute of Environment, University of Tabriz, Tabriz, Iran.
[2] Department of Genetics and Bioengineering, İstanbul Bilgi University, Eyüpsultan, 34060 Istanbul, Turkey.
[3] Department of Biomedical Engineering, Bahcesehir University, Besiktas, 34353 Istanbul Turkey.
[4] Department of Genetics and Bioengineering, Nişantaşı University, Maslak, 34469, Istanbul, Turkey.
* Corresponding author: ashahi@tabrizu.ac.ir

kitchen waste and biodegradable plastics are some examples of biodegradable wastes which can be broken down using composting, aerobic and anaerobic digestion or other similar biological processes. Although biodegradation wastes mostly belong to organic materials originating from living organisms, some inorganic materials can also be broken down so that biodegradation as a key process attenuates the inorganic compounds as hazardous waste (Speight 2017).

Biological processes play a main role in the removal of contaminants from the environment. Microorganisms have developed different removal strategies such as biosorption, bioaccumulation, biotransformation and biodegradation which have the most important role in the bioremediation of contaminated sites (Dixit et al. 2015).

Microbial strains, because of their catabolic versatility, degrade/convert different kinds of pollutants (Khetan 2014). They are able to perform various biochemical cycles, have high metabolic and degradative capabilities, and can biodegrade different wastes in many diverse types of environments (Singh 2013). Inclusive microbial catabolic abilities enable them to degrade, transform or accumulate various types of environmental pollutants such as polycyclic aromatic hydrocarbons (PAHs), petroleum hydrocarbons, pharmaceutical substances, polychlorinated biphenyls (PCBs), radionuclides and heavy metals in large quantities (Joutey et al. 2013).

This chapter explains different strategies which are used by microbes to remove contaminants. Biosorption as physiochemical and passive binding of contaminants into cellular structure, bioaccumulation as gradual metabolic-dependent absorption of contaminant in microorganism, biotransformation as the enzymatic conversion of contaminant in to less toxic compounds and biodegradation as decomposition of organic compounds which leads to produce simple inorganic compounds in case of complete mineralization, are described in detail.

Biosorption

Biosorption is a technique that includes adsorption, absorption, ion exchange, precipitation and surface complexation for removal of contaminants, especially those that are not biodegradable such as metals from the polluted sites, radionuclides including actinides and lanthanides and some biodegradable and recalcitrant organics as well (Vijayaraghavan and Yun 2008, Gadd 2009). Heavy metals as the main group of inorganic contaminants which enter the environment from different sources include industrial activities, agricultural chemicals, car exhausts and petroleum refining process. These sorts of contaminants are hazardous to the environment, pose human health risks and are harmful to other living organisms (El-Naggar et al. 2018). Bacteria, fungi and microalgae are able to remove metals by biosorption (Gadd 2009). In general, all kinds of biomaterial have biosorption removal capability; however, because of their cell wall structure and nature of the cell wall constituents, bacteria have the highest biosorption (Vijayaraghavan and Yun 2008). Both living and dead cells do biosorption (Gadd 2009). The degree of biosorption is mostly determined by the bacterial genus type and also the kind of metal ions. Biosorption degree mostly depends on types of contaminants, and kinds of bacteria due to different cell wall compositions (Michalak et al. 2013). Both gram negative and gram positive bacteria, because of having overall negative charge and anionic character on their cell wall compounds (i.e., tichoic acids of gram positive bacteria and lipopolysaccharides of gram negative bacteria), are capable of adsorbing metal ions (Sherbert 1978, Johnson et al. 2007). In dead cells, the extracellular functional groups such as carboxyl, hydroxyl and amine have very important roles in biosorption (Bilal et al. 2018). Amine is an important functional group which removes the metal ions by chelating of metal ions and by adsorbing anionic metal species via hydrogen bonding and electrostatic interactions, as well (Navarro et al. 1998, Mohseni et al. 2019). Extracellular polysaccharides can also bind to metal ions, but these compounds may be washed out or removed during waste treatment process (Gupta and Diwan 2017). Biosorption is affected by many environmental factors such as temperature, pH, ionic strength, etc. (Bilal et al. 2018, El-Naggar 2018). Metal ions hydrolysis occurs as pH increases, then chemical species of metal changes with any variation in pH which leads to variable charges and different absorbability of metal ions (Barakat 2008). Biosorption of Cr by *Pseudomonas* spp. (Ozturk et al. 2012) and U by *Saccharomyces cerevisiae* (Liu et al. 2010) are some examples of metal biosorption by bacteria.

Biosorption also involves the sorption of organic pollutants on biological matrix (Sun et al. 2016). Many organic compounds in the environment are degraded by biological process via microbial communities and such biodegradation activities are the fundamentals of biological treatment methods (Gadd 2009). However, some organics are recalcitrant to biodegradation or some biodegradation by-product may be more toxic (Jiang et al. 2016). In such cases, biosorption of contaminant would be a good alternative for waste treatment. Phenolic compounds, pesticides and dyes are some reported organics which have received special attention (Juhasz et al. 2002, Kumar et al. 2014, Argun et al. 2017). It is well known that carbon substrates stimulate the growth of cell wall. Surfactant materials are the other elements which increase the solubility of organic compounds and change the interfacial tension of the cell wall matrix. These factors then modify cell wall and affect biosorption (Vijayaraghavan and Yun 2008). Metabolic activities of living cells may also affect biosorption process due to changes in bioavailability and chemical speciation (Haws et al. 2006). As an instance for biosorption of organic compounds, Sun et al. (2016) indicate that 2, 2, 4, 4-tetrabromodiphenyl ether (BDE-47), as a ubiquitous environmental contaminant, is removed by heat killed *Pseudomonas stutzier* KS0013; however, the removal ability of alive cells was significantly larger than that of killed ones due to metabolically mediated biosorption. Biosorption may also promote the transfer of organics into the bacteria and increase the biodegradation rate of the contaminant. In removing of polychlorinated biphenyl (PCB) by *Ensifer adhaerens,* biodegradation increased with time with the help of biosorption process (Xu et al. 2016); polychlorinated dibenzofurans (PCDFs) by *Bacillus pumilus* (Hong et al. 2000), pyrene (0.41 mg/l) by *Brevibacillus brevis* (Liao et al. 2015), and polychlorinated biphenyl (PCB) by *Ensifer adhaerens* (Xu et al. 2016) are some examples of organic compounds biosorption.

Bioaccumulation

Microorganisms are able to accumulate metals and non-metals which allow them to survive and adapt to the contaminant's site (Ayangbenro and Babalola 2017). Bioaccumulation is a metabolism dependent mechanism, while biosorption is almost always a metabolic independent process which can be also done by nonviable microbes (Juwarkar and Yadav 2010). Regarding heavy metals, biosorption has some advantages over bioaccumulation. The recovery of accumulated metal is achieved only by destruction of the biomass while in biosorption, the desorption process can be done using simple physical methods and there is no need to damage the microbe (Pal et al. 2010). Through the bioaccumulation, metals transport and are stored in the cytoplasm by metabolic mechanisms. Microorganisms also have some intracellular detoxification strategies such as sequestration by small peptides; for example, Cd^{2+} is detoxified by binding into small peptides, which is followed by co-precipitation with inorganic sulfide and a nontoxic metal sulfide cluster is produced (Carney et al. 2007). Various peptides in bacteria have been reported for heavy metal accumulation. These peptides, because of having metal binding amino acids such as cysteine and histidine residues, enhance the accumulation process (Mejáre and Bülow 2001). Metal binding proteins such as metallothioneins (MTs), and cysteine-rich proteins with low molecular mass, are also able to bind mono- and divalent metal ions (Freisinger and Vasak 2013). Some bacterial strains such as *Synechococcus* sp., *Pseudomonas putida* and *Thiobacillus* sp. produce metal binding proteins for intracellular chelation of toxic metal to accumulate the metal, to make them inactive metabolically and to increase the resistance of bacteria by a subsequent exportation (Yoshida et al. 1993, Olafson et al. 1988, Highman et al. 1984, Mejáre and Bülow 2001, Mishra and Malik 2013). For enhancing the metal accumulation, scientists have expressed various MTs from different eukaryotic sources in bacteria. Intracellular expression of MTs have shown problems such as short half-time and low stability because of high cysteine content and the probable interference with redox pathways in the cell (Mejáre and Bülow 2001). One way to overcome this problem is to express the metal binding proteins on the cell surface. This approach increased the capability of metal binding protein and is reported as a successful technique in metal accumulation especially for cadmium (Li and Tao 2013). The expression of metal binding peptides also increases the metal binding capacity of bacteria. Metal binding peptides have higher specificity and selectivity for a specific metal than metal binding proteins (DeSilva et al. 2002).

Phytochelatins (PCs) are another metal-binding complex which cause sequestering of the toxic ions and increase heavy metal tolerance in plants and some yeasts and microorganisms (Cobbett 2000). PCs are a kind of cysteine-rich thiol-reactive peptides, whose synthesis is stimulated when cell is exposed to metal ions such as Cu^{2+}, Pb^{2+}, Ag^+ and Zn^{2+}. The produced PCs are then able to bind to different types of metals and metalloids (Inouhe 2005). Glutathione (GSH) is the direct substrate for PCs synthesis. There are few reports of PCs production by prokaryotes. Most prokaryotes synthesize glutathione and glutathione-related peptides but they usually don't have PCs gene. *Alr0975* gene, encoding a PC-like gene, has been identified from *Nostoc* sp. PCC 7120 (Tsuji et al. 2004). *Pseudoalteromonas citrea* and *Marinobacter* sp. are two other examples which showed resistance to Cd by a PCs-mediated system (Ivanova et al. 2002).

Polyphosphate production by some bacteria such as *Citrobacter freundii* to P removal from wastewater is another strategy for detoxification. Wang et al. (2018) reported that the polyphosphate accumulated in an engineered *Citrobacter freundii* (CPP) could reach up to 12.7% of dry weight. The role of polyphosphate bodies in effective sequestration of Mn, Al, Pb, and Zn in *Plectonema boryanum* has been reported by Torres et al. (1998).

Bioaccumulation due to nutrients and cultivation requirement is not an economic process compared to biosorption (Mishra and Malik 2013). The process is also time consuming and slower than biosorption. Bioaccumulation, therefore, would be appropriate only when the strains show an outstanding resistance against the contaminant and can tolerate a wide range of environmental factors such as pH, contaminant concentration, bioavailability, and temperature (Mishra and Malik 2013).

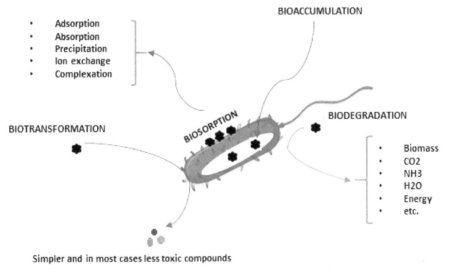

Fig. 1. Microbe-contaminant interactions during bioremediation process.

Biotransformation

Biotransformation is defined as the conversion of chemical compounds not used as nutrients to different products with different applications (Borges et al. 2007). Compared with biodegradation, which refers to the breaking down of complex compound into simpler and in most cases less toxic ones, biotransformation modifies a toxic compound by changing its structure and allows to have new products and improves the production of already known chemicals (Giri et al. 2001). The biotransformation process is catalyzed by microbial enzyme systems in individual cells and extracted enzymes as well (Doble et al. 2014). In recent years, researchers have shown more interest in biotransformation as an effective technique for remediation of contaminants. Most living organisms are able to detoxify complex pollutants; however, because of high growth rate, wide metabolic diversity and their ability in horizontal gene transfer, bacteria have the main role in this process (Shahi et al. 2017, Singh et al. 2017). Microbial strains, based on their diverse catabolic potential, are utilized for the remediation of vast range of pollutants such as polyaromatic hydrocarbons (PAHs), pharmaceutical compounds, petroleum hydrocarbons, polychlorinated biphenyls (PCBs), etc. Different genera from both aerobic and anaerobic bacteria have been reported for their significant contribution to the biotransformation of xenobiotic compounds. *Pseudomonas, Bacillus, Rhodococcus, Micrococcus, Syntrophobacter, Desulfovibrio* are some repeated examples (Chowdhury et al. 2008).

The use of isolated enzymes in various types of industry and scientific researches is increasing rapidly because of some characteristics such as low energy input, eco-friendliness and cost effectiveness (Singh et al. 2016). Microbial enzymes are able to degrade and transform toxic compounds; thus, there are also plenty of works in biotransformation of toxic contaminants using isolated enzymes. However, it has some disadvantages when using enzyme in industry or bioremediation. High cost of purification, providing the necessary cofactors which are needed during process and recycling of derived molecules, are some of the drawbacks of using isolated enzymes (Homaei et al. 2013, Singh et al. 2016). Furthermore, the activity of enzymes is highly sensitive to any change in temperature and pH. They are also very sensitive to substrate and product inhibition (Bisswanger 2019).

Utilization of microorganism in biotransformation of toxic compounds is increasing. Because of some advantages such as having multi-enzymatic system and necessary cofactors, economic friendly, etc., in most cases researchers prefer to use microbe in the biotransformation research. Furthermore, higher growth rate of microbial cells, higher metabolism rate in microbes, easy to provide a sterile condition and high surface to volume ratio are some characters that make microbes as ideal choices for biotransformation (Hegazy et al. 2015).

The selection of suitable microbial strain or consortia and optimizing the operational conditions are crucial to have a successful biotransformation process. Both in microbial and enzymatic biotransformation undesired chemical may be produced which should be considered (Giri et al. 2001).

Microbial transformation can be used for the bioconversion of different chemicals. Steroid and non-steroid compounds, pesticides, antibiotics, petroleum hydrocarbons, BTX (benzene, toluene, xylene), etc., are among the frequently studied compounds.

Steroids are a large group of organic compound that include adrenal-cortical and sex hormones, bile salts, sapogenins, insect molting hormones, alkaloids, etc., which are composed of four carbon rings core structure (Sultan and Reza 2015). Steroid biotransformation mostly refers to synthesis of new steroids as drugs and hormones. For instance, androstenedione is an intermediate steroid which is necessary for the production of steroid drugs. *Arthrobacter, Brevibacterium, Bacillus, Streptomyces* and *Pseudomonas* are known to transfer phytosterol to androstenedione (Rokade et al. 2018). *Corynebacterium* sp., *Nocardia* spp., and *Pseudomonas* sp. are some examples of bacterial genera which have been used in biotransformation production process. Steroid compounds become serious problems even at very low levels when they enter the environment because of their interference with the endocrine system of humans and animals. Several studies have investigated the occurrence of steroids in groundwater, surface water, wastewater and manure (Ting and Praveena 2017). Pharmaceutical industry, municipal and hospital wastewater are the main sources of steroids. Steroids are chemically stable compounds which are found in both conjugated and free forms. The conjugated form itself is readily converted to the free form (Ting and Praveena 2017). Various studies on the biodegradation of steroids showed that the biodegradation rates are varied and depend on operating condition. Non-steroid compounds and antibiotics are also produced by microbial transformation. Dihydroxyacetone, prostaglandins, L-ascorbic acid, etc., can be produced by microbial biotransformation (Roberts et al. 1995). Two new actinomycins are produced by *Streptomyces parvulus* by replacing proline with cis-4-methylproline (Smitha et al. 2017).

Petroleum is the main source of energy for industry and domestic uses. Huge amount of petroleum hydrocarbons enter the environment through different extraction, transportation and refining processes (Rahman et al. 2003). Petroleum waste contains toxic, carcinogenic and mutagenic compounds which cause health problems to humans and ecosystems (Shahi et al. 2016b). Different chemical and physical methods are used to remove the petroleum hydrocarbons from the environment but these methods cannot completely solve the contamination problem and in many cases, increase the toxicity of compounds (Shahi et al. 2016a). Bioremediation with microbial strains is an environmentally friendly method in biodegradation of petroleum derived waste. *Pseudomonas* sp., *Streptococcus* sp., *Desulfitobacterium* sp., *Clostridium* sp., *Kosmotoga* are some active genera in biodegradation of petroleum hydrocarbons (Shahi et al. 2016b).

Biodegradation

Principles of biodegradation

Biodegradation is the biologically catalyzed process for the breaking down of chemical compounds by microorganisms (Joutey et al. 2013). Biodegradation refers to the ability of microorganisms to degrade and recycle environmental contaminants (Alvarez and Illman 2005). Indigenous microorganisms (i.e., innate microbial consortium that inhabits the environment) often biodegrade different types of organic compounds to obtain energy and synthesis of biomass for their growth and reproduction (Kumar and Gopal 2015).

Different aspects of microbial biodegradation have evolved various terms such as bioremediation, bioattenuation, biostimulation and bioaugmentation in this field. Bioremediation is a process that uses microorganisms and plants to biodegrade contaminants in contaminated environments (Loss and Yu 2018). When the bioremediation occurs by native microorganisms without human intervention and only with monitoring of any changes in pollution concentrations, the process is called natural attenuation or bioattenuation. Natural attenuation is the simplest form of bioremediation (Alvarez and Illman 2005). In most environments, the indigenous microbes have adapted to the biodegradation of the contaminant because of long term exposure, or are naturally selected to such a site (Hazen 2010). Biostimulation refers to the enhancement of biodegradation rate in the contaminated site by adding of necessary nutrients such as nitrogen and phosphorus, oxygen and other electron acceptors (Shahi et al. 2016a). Like in bioattenuation, biostimulation occurs by indigenous microbes, therefore in addition to the presence of necessary microorganisms, the bioremediation site should be capable of being altered for the anticipated bioremediation effect. For a successful biostimulation, knowing about the biogeochemistry of site and finding the limiting conditions of native microbes are the most critical points (Hazen 2010). Biodegradation rate is highly variable for different environments. Microbial diversity and environmental conditions determine the biodegradation rate and biodegradation potential as well (Alvarez and Illman 2005). Sometimes the biodegradation process in the environment proceeds under favorable condition without any human intervention; however, usually the biodegradation in a polluted site needs to be enhanced by altering the site's physical and chemical characteristics or increasing the microbial richness by adding efficient microbial strains.

When bacteria is grown in a pure culture (i.e., only one strain of bacterium in a medium), the number of bacterial cells will increase dramatically. Under optimal environmental parameters and by supplying all required nutrients, the cell population increment can be measured over time and a growth curve can be developed (Maier 2009). Growth curve and analyzing of microbial culture is an important task during biodegradation process, for example, to interpret the growth rate, to optimize growth condition and to increase the biodegradation rate of a given contaminant (Hall et al. 2014, Berbert-Molina et al. 2008). Several distinct phases include the lag phase, the exponential or log phase, the stationary phase, and the death phase, which are detectable in a growth curve (Fig. 2) (Wang et al. 2015). Each phase indicates a discrete growth period with its own growth rate which shows distinctive physiological changes in the cell culture. Lag phase is a period during which the growth rate of microbe due to the physiological adaptation and low extracellular enzymes, is so slow or essentially zero (Maier 2009, Wang et al. 2015). Depending on the type of medium and initial size of inoculum, the lag phase can last from few minutes up to several hours. The exponential phase is a period in which the growth rate is optimal. During this phase, the cell numbers double at mean generation time which is the average interval between the birth of a microbe and the birth of its offspring. The most common way to express the cell increment in the culture is geometric progression in which each cell division results in a doubling of the cell number 2^0, 2^1, 2^2, 2^3, 2^4 ...,2^n. After n divisions, there are 2^n cells. The number of cells at time t is calculated according to the formula given below:

$$N_t = N_0 \cdot 2^n \qquad\qquad\text{(Eq. 1)}$$

where:

N_t = cell number at the time t
N_0 = initial cell number
n = generation number

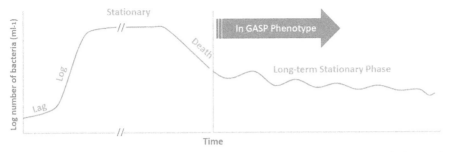

Fig. 2. Five phases of bacterial growth curve: lag phase, log phase, stationary phase, death phase and long-term stationary phase which may take months or even years for a bacterial strain in culture condition (modified from Finkel 2006 and Navarro-Llorens et al. 2010).

Stationary phase is characterized by a state of no net growth. In this phase, growth (cell division) is balanced by the number of cell dying. Reduction in growth rate is mostly because of the fact that the carbon and energy sources or necessary nutrients have been ended. Increasing of waste products up to the toxic or inhibiting point may also cause the reduction in growth rate in stationary phase (Maier 2009). The small amount of growth in Fig. 2 is mostly because of endogenous metabolism which occurs via cell endogenous sources or carbon and nutrients obtained from lysing of dead cells in the obscene of necessary compounds (Dawes and Ribbons 2003). Due to the reduction of some compounds, growth process in stationary phase is unbalanced. Because of this, cells harvested from stationary phase are usually smaller than cells in exponential phase and hence inappropriate for use as inoculum in biodegradation studies (Jaishankar and Srivastava 2017). As carbon source and other essential compounds become less available and with increasing of waste products, the number of living cells decrease exponentially and the death phase begins. Through the death phase, death rate surpasses growth rate leading to a net loss of viable cells. Lysing of dying cells in this phase produces nutrients available for other bacteria and gives opportunity to spore producing bacteria to live long enough for spore production (Navarro-Llorens et al. 2010). Spore forming helps bacteria to survive unfavorable condition of the death phase.

Zambrano et al. (1993) showed that *rpoS* mutant *Escherichia coli* compete and survive in stationary phase. This strain, which unlike unchanged wild-type cells, presented growth advantage in stationary phase and replaced the original population under the strong selective force of starvation in *E. coli* culture. Phenotypes, which show growth advantage in stationary phase (i.e., GASP phenotypes), showed a fifth phase which was named Long-term stationary phase by Finkel (2006) (Fig. 2). This period is a dynamic phase in which the birth and death rates are balanced. In natural environment and in contaminated sites as well, bacteria live in situations similar to long-term stationary phase and are exposed to stressful conditions (Finkel 2006, Helmus et al. 2012). In such circumstances, bacteria express different genes and alternative metabolic pathways which are necessary for survival. Therefore, the selection of GASP phenotype by stressful environment will be definitive (Finkel 2006).

There are two approaches to study the bacterial growth and biodegradation manner of a given pollutant under controlled conditions. Batch culture is a small-scale laboratory experiment in which a defined volume of medium in a small volume flask is inoculated with single microorganism or a group of microorganisms (Rouf et al. 2017). Batch culture is a closed culture system and a fixed amount of substrate, which could be a contaminant, is added at the beginning. During the operation process, the amount of nutrients and carbon source are depleted and waste materials are accumulated. In continuous culture system, the amount of available substrate remains the same by a continuous influx of growth medium and substrate into the system (Maier 2009). This system can be operated for long time and a continuous effluent stream containing waste products, metabolites, unused materials and also microbial cell will be withdrawn from the system. Continuous culture system is done in a growth vessel which is called chemostat or bioreactor. In a bioreactor the flow rate, steady concentration of substrate, pH, oxygen level and temperature can be controlled and the operation condition can also be optimized during the operation (Doran 2013).

For describing entire growth curve and biodegradation process in batch system, some mathematical models have been developed. Models combine theories, that offer the structure of the model and experimental observations which suggests the numerical values of constants and coefficients (Maier 2009). Specifying of the model is necessary for a biological process because of wide range of environmental factors and complexity of cellular system (Doran 2013). A frequently used mathematical model is Monod model. Monod model has been proposed by Jacques Monod in 1940s for growth of microorganisms. It expresses the rate of substrate consumption in terms of cell number.

$$\frac{dx}{dt} = \mu x = \frac{\mu_{max} S X}{K_S + S} \qquad (Eq. 2)$$

where X is the number of cells (mass/volume), t is time, μ is the specific growth rate (time^{-1}), μ_{max} is the maximum specific growth rate (time^{-1}), S is the substrate concentration (mass/volume) and K_S is the half-saturation constant (mass/volume) which is the value of S when $\mu/\mu_{max} = 0.5$. Both μ_{max} and K_S are inherent physiological properties of microbe and depend on the substrate that is used. In Monod model, it is assumed that nutrients except substrate is not limiting and toxic waste which could inhibit the biological process is not produced.

Based on the Monod equation, when S is very higher than K_S and the substrate is high, growth will be at its maximum rate. This situation occurs in the initial stage of batch culture when substrate and nutrient are in high level, and during all system operation period in continues culture which the amount of nutrient and substrate are almost constant (Flynn et al. 2018). In natural condition, in contaminated soil and aqueous environments where nutrients and substrate are usually limiting, this kind of growth is far to happen (Maier 2009). In low substrate concentration where S is very lower than K_S, the rate of growth will also decrease. This kind of growth is found in batch culture at the end point when the substrate and nutrients have been utilized. This type of growth is also expected in natural environments where the substrate and nutrient concentrations are low. Monod equation can be rewritten in function of substrate utilization. Substrate-to-biomass ratio is an important parameter in Monad modeling. Monod equation in terms of substrate biodegradation and yield coefficient can be rewritten as below:

$$\frac{dS}{st} = -\frac{1}{Y}\frac{\mu_{max}SX}{K_S + S} \qquad \text{(Eq. 3)}$$

where Y is the yield coefficient (mass of cells formed/mass of substrate consumed).

A yield coefficient in a biodegradation system means that the substrate is biodegraded more efficiently.

In continuous systems, dilution rate, which means the rate of nutrient exchange in the bioreactor, is an important parameter in system operation process. Dilution rate is equal to the flow rate divided by the bioreactor volume. When dilution rate is high, it means that the amount of substrate utilization is high and substrate conversion efficiency is low, whereas a low dilution rate means that substrate is efficiently converted (Flynn et al. 2018). In a steady state condition, which means no changes in biomass concentration over time, the highest possible dilution rate (i.e., critical dilution rate) will be obtained (Maier 2009, Flynn et al. 2018). In such condition, the critical dilution rate (D_C) is equal to specific growth rate (μ). Using equation (Eq. 2), the critical dilution rate will be:

$$D_C = \mu_{max}\frac{S}{K_S + S} \qquad \text{(Eq. 4)}$$

where D_C is the critical dilution rate (l.time^{-1}).

When D is higher than μ_{max} the cell removal rate will be faster than growth rate, and thus culture will be washed out (Faschian et al. 2017).

Aerobic biodegradation

Aerobic condition provides an environment under which microorganisms breakdown organic pollutants by aerobic respiration process. Aerobic metabolism is a fast and effective process for the biodegradation of waste materials (Herzyk et al. 2014). Aerobic bacteria using aerobic respiration as an efficient energy source, break down the organic compounds and convert the contaminates and waste inputs into carbon dioxide, water and energy. Nutrients, ammonium, humus and stable forms of organic matters are also produced through aerobic biodegradation process. Aerobic biodegradation process has been used for some waste management purposes such as municipal and industrial wastewater treatment, stabilization of organic matter through composting, oily sludge biodegradation, biodegradation of toxic materials such as polymers, etc. Some xenobiotic compounds such as toluene, benzene, phenol, naphthalene and so on, the chemicals which are not naturally produced and are foreign to the biosphere, are also degraded by aerobic bacteria. These bacteria are able to produce enzymes which degrade xenobiotics to nontoxic compounds (Agrawal and Dixit 2005). Through the biological biodegradation of organic compounds, more than half of the materials convert to the energy and the remaining section is used for producing of new cells and biomass which may be later utilized for energy production through endogenous respiration (Metcalf and Eddy 2003).

The stoichiometry of aerobic oxidation for growth purpose is given below (Orhon and Artan 1997):

$$8CH_2O + 3O_2 + NH_3 \longrightarrow C_5H_7NO_2 + 3CO_2 + 6H_2O$$

where CH_2O is substrate and $C_5H_7NO_2$ is a general formula which represents the biomass.

The equation can be used to calculate the amount of oxygen and also nitrogen required for biodegradation of a particular substrate. Remediation of contaminated sites, treatment of municipal, agricultural, pharmaceutical and industrial wastewater and so on are some examples in which the above mentioned stoichiometry equation is used.

Endogenous metabolism takes place when extra cellular organic substrate is almost removed. In this condition, the microbial population is in its maximum size and the cellular matter is the main substrate source. In this phase, endogenous respiration as the second phase of oxidation will be dominant (Sperling 2007).

The equation below can be used for stoichiometry of endogenous respiration (Orhon and Artan 1997):

$$C_5H_7NO_2 + 5O_2 \longrightarrow 5CO_2 + 2H_2O + NH_3$$

Endogenous decay is an important parameter in biodegradation processes especially in wastewater treatment process. Sludge production is an unavoidable consequence of wastewater treatment systems. It is a secondary solid waste that needs to be disposed and is a real challenge in wastewater treatment process. The lower endogenous decay rate leads to increase in the net production of sludge when operation of wastewater treatment plant (Water Environmental Federation 2010).

In biodegradation process, when an organic substrate is converted to its inorganic constituent and becomes harmless, it is termed mineralization. In other words, mineralization is a type of biodegradation which refers to total degradation of organic matter (Martin 1999). As some pollutants could not be utilized by bacteria as sole carbon source, these sort of contaminants are not degraded completely and could just be transformed by some microbes as non-growth materials, a process termed cometabolism. In this type of biodegradation, transformation process is usually catalyzed by non-specific enzymes of other metabolic pathways (Martin 1999). In pure culture, cometabolism of a contaminant results in only partial degradation. In this case, no bacterial growth occurs and by-products accumulate in the medium. In contaminated environment, a cometabolism by-product of one bacteria can be metabolized by other one and can lead to complete mineralization in some cases (Alvarez and Illman 2005). In natural environments, cometabolism of many contaminants occurs very slowly since the microbial population responsible for biodegradation does not have enough size; however, wild type microorganisms can be used for construction of microbial strains with high biodegradation potential of recalcitrant organics (Janke and Fritsche 1985).

Anaerobic biodegradation

Biodegradable contaminants can also be degraded in the absence of oxygen via anaerobic biodegradation. Anaerobic degradation is an effective process to treat high strength organics such as industrial wastes and wastewater sludge (Ersahin et al. 2011). The process has been broadly used in the last few decades because of its several advantages over aerobic treatment. Higher organic loading rate, lower production of excess sludge, lower nutrient requirement, no need for oxygenation and methane rich biogas production through the process are some advantages of anaerobic biodegradation (Muralikrishna and Manickam 2017). Anaerobic digestion is a series of processes which starts with hydrolysis. Through this process, the organic compounds are hydrolyzed to be available for other bacteria (Singh 2013). Acidogenic bacteria then convert the simpler compounds to short-chain organic acids. Some intermediate compounds such as alcohol, carbon dioxide, aldehydes, and hydrogen are also produced during this step (Khan et al. 2006). Acidogenic bacteria mostly belong to fermentative genera such as *Clostridium, Bacteroides, Pseudomonas, Streptococcus, Bacillus,* etc. Acidogens are facultative or strict anaerobes. The facultative members protect oxygen sensitive methanogens by utilizing trace amounts of oxygen in the environment (Anderson et al. 2003).

In the third phase, acetogenic bacteria transform the volatile organic compounds from acidogenesis phase into acetate, carbon dioxide and hydrogen. Two groups of bacteria carry out this stage. Obligate hydrogen producing acetogens are the first group which convert fatty acids and alcohols to acetic acid, carbon dioxide and hydrogen. The second group are homoacetogenic bacteria. Homoacetogens are strictly anaerobic bacteria which reduce carbon dioxide to acetic acid using hydrogen as electron acceptor. *Clostridium, Acetobacterium, Butribacterium* and *Acetoanaerobium* are some examples of homoacetogen bacteria (Anderson et al. 2003). In final step, methane is produced through methanogenesis. Methanogens produce methane in two ways. Acetoclastic methanogens cleavage acetic acid to methane and carbon dioxide, while hydrogenotrophic methanogens reduce carbon dioxide with hydrogen. *Methanosaeta* and *Methanosarcina* are two methanogenic genera which utilize acetate to produce methane. Genome sequencing showed that the majority of pathway core steps are the same in both genera, but there is a remarkable difference in energy conservation and electron transfer chain (Smith and Ingram-Smith 2007). As methanogens are very sensitive to pH changes, pH range of 6.5–8.0 is the most suitable, and any pH changes in the environment which may arise from products of acidogenic and acetogenic bacteria should be buffered (Anderson et al. 2003). Hydrogen-utilizing are second group of methanogens which reduce carbon dioxide using hydrogen to methane. These bacteria can also reduce some other substrates such as formate, methylamines and fomate. Hydrogen-utilizing methanogens are more tolerant to environmental changes when compared with acetoclastic methanogens (Xing et al. 1997).

Stoichiometric equation to calculate the theoretical anaerobic products is very similar to that for aerobic degradation (Tezel et al. 2011).

$$C_cH_hO_oN_nS_s + yH_2O \rightarrow xCH_4 + nNH_3 + sH_2S + (c-x)\,CO_2$$
$$x = 1/8(4c - h - 2o - 3n - 2s)$$
$$y = 1/4(4c - h - 2o + 3n + 3s)$$

The methane ratio of gas mixture in anaerobic process is determined by the compounds which are more reduced. Substrates such as methanol and lipid produce more amount of methane when they degrade anaerobically (Maier 2009).

Table 1 summarizes bacterial strains which have been reported for their abilities in biodegradation of some toxic chemicals that affect human health. Table has been arranged for persistent organic pollutants. Persistent organic pollutants (POPs) are

Table 1. Persistent organic pollutants biodegradation by various bacterial strains.

Compound	Bacterial strain	Reference
Dieldrin	*Pseudonocardia* sp. *Burkholderia* sp. *Cupriavidus* sp.	Sakakibara et al. 2011 Matsumoto et al. 2008
Aldrin	*Pseudomonas fluorescens*	Bandala et al. 2006
Heptachlor	*Pseudomonas fluorescens*	Bandala et al. 2006
Chlordane	*Streptomyces* sp.	Cuozzo et al. 2012
Chlordecone	*Citrobacter* sp.	Chaussonnerie et al. 2016
Decabromodiphenyl ether	*Stenotrophomonas* sp.	Wu et al. 2018
Endrin	*Burkholderia* sp. *Cupriavidus* sp.	Matsumoto et al. 2008
Hexabromocyclododecane	*Sphingobium chinhatense*	Heeb et al. 2017
Hexachlorobenzene	*Nocardioides* sp.	Takagi et al. 2009
Hexachlorobutadiene	*Serratia marcescens* sp.	Li et al. 2008
Alpha hexachlorocyclohexane	*Pandoraea* sp.	Okeke et al. 2002
Beta hexachlorocyclohexane	*Sphingobium japonicum* *Dehalobacter* sp.	van Doesburg et al. 2005 Ito et al. 2007
Gamma hexachlorocyclohexane (Lindane)	*Chromohalobacter* sp.	Bajaj et al. 2017
Pentachlorobenzene	*Dehalococcoides* sp.	Jayachandran et al. 2003
Pentachlorophenol	*Desulfitobacterium hafniense*	Villemur 2013
Polychlorinated biphenyls (PCB)	*Dehalococcoides mccartyi*	Chen and He 2018
Short-chain chlorinated Paraffins (SCCPs)	*Pseudomonas* sp.	Lu 2012
Technical endosulfan and its related isomers	*Pusillimonas* sp. *Bordetella petrii*	Kong et al. 2018
Tetrabromodiphenyl ether	*Pseudomonas stutzeri*	Zhang et al. 2013
Pentabromodiphenyl ether	*Nitrosopumilus* *Nitrososphaera* *Methanomethylovorans*	Yan et al. 2017
Toxaphene	*Enterobacter cloacae*	Lacayo-Romero et al. 2005
DDT	*Stenotrophomonas* sp.	Pan et al. 2016
Perfluorooctane sulfonic acid	*Pseudomonas aeruginosa*	Kwon et al. 2014
Polychlorinated dibenzo-*p*-dioxins (PCDD)	*Pseudomonas mendocina*	Lin et al. 2014
Polychlorinated dibenzofurans	*Terrabacter* sp. *Pseudomonas* sp.	Habe et al. 2001

organic substances that are resistant to chemical, physical and biological degradation process. POPs are discussed and listed in Stockholm Convention on Persistent Organic Pollutants in 2001.

Conclusion

Modern lifestyles and industrial activities generate huge amount of more complex wastes. Microbes have very important role in waste treatment because of their physiological and metabolic versatility, environmental adaptation and biodiversity. Microbial strains are able to remove, transform, immobilize, detoxify and degrade various types of contaminants. Microbial cell wall structure and composition has given them a great biosorption ability, so that they can remove different inorganic and organic contaminants by adsorption into the cell wall matrix. Microbes can also accumulate metal and non-metal contaminants using small peptides, metal binding proteins, phytochelatins, etc. Bioaccumulation, unlike biosorption, is a metabolic-dependent process and has advantage in storing of chemicals which are not able to degrade directly and may affect individual health. Biomineralization and biotransformation are other strategies to completely degrade contaminants or convert them to simpler and less toxic ones. Microbes using various enzymatic systems can degrade vast variety of contaminants such as pesticides, pharmaceuticals and polyaromatic hydrocarbons in aerobic and anaerobic environments. Behavior of a microbe or microbial community in the face of contaminant, mechanisms which they use to remove the contaminant and different strategies which they utilize to increase their capabilities to successfully degrade contaminants are some issues which researchers are actively investigating.

References

Agrawal, N. and Dixit, A. K. 2015. An environmental cleanup strategy-microbial transformation of Xenobiotic compounds. International Journal of Current Microbiology and Applied Sciences 4(4): 429–461.

Alvarez, P. J. J. and Illman, W. A. 2005. Biodegradation principles. *In*: Bioremediation and Natural Attenuation: Process Fundamentals and Mathematical Models. John Wiley & Sons, New Jersey.

Anderson, K., Sallis, P. and Uyanik, S. 2003. Anaerobic treatment processes. *In:* Mara, D. and Horan, N. J. [eds.]. The Handbook of Water and Wastewater Microbiology. Academic Press.

Argun, Y. A., Karacali, A., Calisir, U., Kilinc, N. and Irak, H. 2017. Biosorption method and biosorbents for dye removal from industrial wastewater: a review. International Journal of Advanced Research 5(8): 707–714. doi:10.21474/ijar01/5110.

Ayangbenro, A. S. and Babalola, O. O. 2017. A New Strategy for Heavy Metal Polluted Environments: A Review of Microbial Biosorbents. doi:10.3390/ijerph14010094.

Bajaj, S., Sagar, S., Khare S. and Singh, D. K. 2017. Biodegradation of γ-hexachlorocyclohexane (lindane) by halophilic bacterium Chromohalobacter sp. LD2 isolated from HCH dumpsite. International Biodeterioration & Biodegradation 122: 23–28. doi:10.1016/j.ibiod.2017.04.014.

Bandala, E., Andres-Octaviano, J., Pastrana, P. and Torres, L. 2006. Removal of aldrin, dieldrin, heptachlor, and heptachlor epoxide using activated carbon and/or *Pseudomonas fluorescens* free cell cultures. Journal of Environmental Science and Health, Part B Pestic. Food Contam. Agric. Wastes. 41(5): 553–569. doi:10.1080/03601230600701700.

Barakat, M. A. 2008. Adsorption of heavy metals from aqueous solutions on synthetic zeolite. Research Journal of Environmental Sciences 2(1): 13–22.

Barbuta, M., Bucur, R. D., Cimpeanu, S. M., Paraschiv, G. and Bucur, D. 2015. Wastes in building materials industry. *In*: Pilipavičius, V. [ed.]. Agroecology. IntechOpen, DOI: 10.5772/59933.

Berbert-Molina, M. A., Prata, A. M. R., Pessanha, L. G. and Silveira, M. M. 2008. Kinetics of *Bacillus thuringiensis* var. *israelensis* growth on high glucose concentrations. Journal of Industrial Microbiology & Biotechnology 35(11): 1397–1404. doi:10.1007/s10295-008-0439-1.

Bidlingmaier, W. and Denecke, M. 2016. Microbial fundamentals. *In:* Biological Waste Treatment Technologies.

Bilal, M., Rasheed, T., Sosa-Hernández, J. E., Raza, A., Nabeel, F. and Iqbal, H. M. N. 2018. Biosorption: An interplay between marine algae and potentially toxic elements—A review. Marine Drugs 16(2): 1–16. doi:10.3390/md16020065.

Bisswanger, H. 2019. Practical Enzymology, Wiley

Borges, K. B., de, W., Borges, S., Pupo, M. T. and Bonato, P. S. 2007. Endophytic fungi as models for the stereoselective biotransformation of thioridazine. Applied Microbiology Biotechnology 77: 669–674.

Carney, C. K., Harry, S. R., Sewell, S. L. and Wright, D. W. 2007. Biomineralization I—Detoxification Biominerals. Biomineralization I 155–185. doi:10.1007/128_050.

Chaussonnerie, S., Saaidi, P. L., Ugarte, E., Barbance, A., Fossey A. Barbe, V. et al. 2016. Microbial degradation of a recalcitrant pesticide: Chlordecone. Frontiers in Microbiology 7(DEC). doi:10.3389/fmicb.2016.02025.

Chen, C. and He, J. 2018. Strategy for the rapid dechlorination of polychlorinated biphenyls (PCBs) by Dehalococcoides mccartyi strains. Environmental Science & Technology 52(23): 13854–13862. doi:10.1021/acs.est.8b03198.

Chowdhury, A., Pradhan, S., Saha, M. and Sanyal, N. 2008. Impact of pesticides on soil microbiological parameters and possible bioremediation strategies. Indian Journal of Microbiology 48(1): 114–127.

Cobbett, C. S. 2000. Phytochelatins and their roles in heavy metal detoxification. Plant Physiology 123(3): 825–832. doi:10.1104/pp.123.3.825.

Cuozzo, S. A., Fuentes, M. S., Bourguignon, N., Benimeli, C. S. and Amoroso, M. J. 2012. Chlordane biodegradation under aerobic conditions by indigenous *Streptomyces* strains. International Biodeterioration and Biodegradation 66(1): 19–24. doi:10.1016/j.ibiod.2011.09.011.

Dawes, E. A. and Ribbons, D. W. 2003. The endogenous metabolism of microorganisms. Annual Review of Microbiology 16(1): 241–264. doi:10.1146/annurev.mi.16.100162.001325.

DeSilva, T. M., Veglia, G., Porcelli, F., Prantner, A. M. and Opella, S. J. 2002. Selectivity in heavy metal-binding to peptides and proteins. Biopolymers 64(4): 189–197. doi:10.1002/bip.10149.

Dixit, R., Wasiullah, Malaviya, D., Pandiyan, K., Singh, U. B., Sahu, A. et al. 2015. Bioremediation of heavy metals from soil and aquatic environment: An overview of principles and criteria of fundamental processes. Sustainability 7(2): 2189–2212. doi:10.3390/su7022189.

Do, K. and Bui, B. 2018. A novel approach of cleaner production concepts to evaluate promising technologies for sludge reduction in wastewater treatment plants. SF Journal of Environmental and Earth Science 1: 1–9.

Doble, M., Kruthiventi, A. K. and Gaikar, V. G. 2014. Biotransformations and Bioprocesses. Taylor & Francis, USA, New York.

Doran, P. M. 2013. Bioprocess Engineering Principles. Academic Press.

El-Naggar, N. E. A., Hamouda, R. A., Mousa, I. E., Abdel-Hamid, M.S. and Rabei, N. H. 2018. Biosorption optimization, characterization, immobilization and application of Gelidium amansii biomass for complete Pb^{2+} removal from aqueous solutions. Scientific Reports 8(1): 1–19. doi:10.1038/s41598-018-31660-7.

Ersahin, M. E., Ozgun, H., Dereli, R. K. and Ozturk, I. 2011. Anaerobic treatment of industrial effluents: An overview of applications. Waste Water Treatment and Reutilization 9–13. doi:10.5772/16032.

Faschian, R., Eren, I., Minden, S. and Pörtner, R. 2017. Evaluation of fixed-bed cultures with immobilized *Lactococcus lactis* ssp. *lactis* on different scales. The Open Biotechnology Journal 11(1): 16–25. doi:10.2174/1874070701711010016.

Finkel, S. E. 2006. Long-term survival during stationary phase: Evolution and the GASP phenotype. Nature Reviews Microbiology 4(2): 113–120. doi:10.1038/nrmicro1340.

Flynn, K. J., Skibinski, D. O. F. and Lindemann, C. 2018. Effects of growth rate, cell size, motion, and elemental stoichiometry on nutrient transport kinetics. PLOS Computational Biology 14(4). https://doi.org/10.1371/journal.pcbi.1006118.

Freisinger, E. and Vasak, M. 2013. Cadmium in metallothioneins. *In:* Sigel, A., Sigel, H. and Sigel, R. K. O. [eds.]. Cadmium: From Toxicity to Essentiality. Springer, Netherlands.

Gadd, G. M. 2009. Biosorption: Critical review of scientific rationale, environmental importance and significance for pollution treatment. Journal of Chemical Technology & Biotechnology 84(1): 13–28. doi:10.1002/jctb.1999.

Giri, A., Dhingra, V., Giri, C. C., Singh, A., Ward, O. P. and Narasu, M. L. 2001. Biotransformations using plant cells, organ, culture and enzyme systems: Current trends and future prospects. Biotechnology Advances 19: 175–199.

Gupta, P. and Diwan, B. 2017. Bacterial exopolysaccharide mediated heavy metal removal: A Review on biosynthesis, mechanism and remediation strategies. Biotechnology Reports 13: 58–71. doi:10.1016/j.btre.2016.12.006.

Habe, H., Chung, J. S., Lee, J. H., Kasuga, K., Yoshida, T., Nojiri, H. et al. 2001. Degradation of chlorinated dibenzofurans and dibenzo-p-dioxins by two types of bacteria having angular dioxygenases with different features. Applied and Environmental Microbiology 67(8): 3610–3617. doi:10.1128/AEM.67.8.3610-3617.2001.

Hall, B. G., Acar, H., Nandipati, A. and Barlow, M. 2014. Growth rates made easy. Molecular Biology and Evolution 31(1): 232–238. doi:10.1093/molbev/mst187.

Haws, N. W., Ball, W. P. and Bouwer, E. J. 2006. Modeling and interpreting bioavailability of organic contaminant mixtures in subsurface environments. Journal of Contaminant Hydrology 82(3-4): 255–292. doi:10.1016/j.jconhyd.2005.10.005.

Hazen, T. C. 2010. Biostimulation. *In:* Timmis, K. N. [ed.]. Handbook of Hydrocarbon and Lipid Microbiology. Springer Berlin Heidelberg.

Heeb, N. V., Grubelnik, A., Geueke, B., Kohler, H. P. E. and Lienemann, P. 2017. Biotransformation of hexabromocyclododecanes with hexachlorocyclohexane-transforming Sphingobium chinhatense strain IP26. Chemosphere 182: 491–500. doi:10.1016/j.chemosphere.2017.05.047.

Hegazy, M. E. F., Mohamed, T. A. and ElShamy, A. I. 2015. Microbial biotransformation as a tool for drug development based on natural products from mevalonic acid pathway: A review. Journal of Advanced Research 6(1): 17–33. doi:10.1016/j.jare.2014.11.009.

Helmus, R. A., Liermann, L. I., Brantley, S. L. and Tien, M. 2012. Growth advantage in stationary-phase (GASP) phenotype in long-term survival strains of *Geobacter sulfurreducens*. FEMS Microbiology Ecology 79(1): 218–228. doi:10.1111/j.1574-6941.2011.01211.x.

Herzyk, A., Maloszewski, P., Qiu, S., Elsner, M. and Griebler, C. 2014. Intrinsic potential for immediate biodegradation of toluene in a pristine, energy-limited aquifer. Biodegradation 25(3): 325–336. doi:10.1007/s10532-013-9663-0.

Higham, D. P., Sadler, P. J. and Scawen, M. D. 1984. Cadmium-resistant *Pseudomonas putida* synthesizes novel cadmium proteins. Science 225: 1043–1046.

Homaei, A. A., Sariri, R., Vianello, F. and Stevanato, R. 2013. Enzyme immobilization: An update. Journal of Chemical Biology 6(4): 185–205. doi:10.1007/s12154-013-0102-9.

Hong, H. B., Hwang, S. H. and Chang, Y. S. 2000. Biosorption of 1,2,3,4-tetrachlorodibenzo-p-dioxin and polychlorinated dibenzofurans by *Bacillus pumilus*. Water Research 34: 349–353.

Inouhe, M. 2005. Phytochelatins. Brazilian Journal of Plant Physiology 17(1): 65–78.

Ito, M., Prokop, Z., Klvaňa, M., Otsubo, Y., Tsuda, M., Damborský, J. et al. 2007. Degradation of β-hexachlorocyclohexane by haloalkane dehalogenase LinB from γ-hexachlorocyclohexane-utilizing bacterium Sphingobium sp. MI1205. Archives of Microbiology 188(4): 313–325. doi:10.1007/s00203-007-0251-8.

Ivanova, E. P., Kurilenko, V. V., Kurilenko, A. V., Gorshkova, N. M., Shubin, F. N., Nicolau, D. V. et al. 2002. Tolerance to cadmium of free-living and associated with marine animals and eelgrass marine gamma-proteobacteria. Current Microbiology 44: 357–362.

Jaishankar, J. and Srivastava, P. 2017. Molecular basis of stationary phase survival and applications. Frontiers in Microbiology 8(OCT): 1–12. doi:10.3389/fmicb.2017.02000.

Janke, D. and Fritsche, W. 1985. Nature and significance of microbial cometabolism of xenobiotics. Journal of Basic Microbiology 25(9): 603–19.

Jayachandran, G., Görisch, H. and Adrian, L. 2003. Dehalorespiration with hexachlorobenzene and pentachlorobenzene by *Dehalococcoides* sp. strain CBDB1. Archives of Microbiology 180(6): 411–416. doi:10.1007/s00203-003-0607-7.

Jiang, Y., Brassington, K. J., Prpich, G., Paton, G. I., Semple, K. T., Pollard, S. J. T. et al. 2016. Insights into the biodegradation of weathered hydrocarbons in contaminated soils by bioaugmentation and nutrient stimulation. Chemosphere. 161: 300–307. doi:10.1016/j.chemosphere.2016.07.032.

Johnson, K. J., Ams, D. A., Wedel, A. N., Szymanowski, J. E. S., Weber, D. L., Schneegurt, M. A., et al. 2007. The impact of metabolic state on Cd adsorption onto bacterial cells. Geobiology 5(3): 211–218. doi:10.1111/j.1472-4669.2007.00111.x.

Joutey, N. T., Bahafid, W., Sayel, H. and El Ghachtouli, N. 2013. Biodegradation: Involved microorganisms and genetically engineered microorganisms. *In*: Chamy, R. and Rosenkranz, F. [eds.]. Biodegradation—Life of Science. IntechOpen, DOI: 10.5772/56194.

Juhasz, A. L, Smith, E., Smith, J. and Naidu., R. 2002. Biosorption of organochlorine pesticides using fungal biomass. Journal of Industrial Microbiology and Biotechnology 29(4): 163–169. doi:10.1038/sj.jim.7000280.

Juwarkar, A. A. and Yadav, S. K. 2010. Bioaccumulation and biotransformation of heavy metals. *In*: Fulekar, M.H. [ed.]. Bioremediation Technology: Recent Advances. Springer, Netherlands.

Khan, M. A., Ngo, H. H., Guo, W. S., Liu, Y. W., Zhou, J. L., Zhang, J. et al. 2016. Comparing the value of bioproducts from different stages of anaerobic membrane bioreactors. Bioresource Technology 214: 816–825. doi:10.1016/j.biortech.2016.05.013.

Khetan S. K. 2014. Endocrine Disruptors in the Environment (1st Edition). Wiley, New Jersey.

Kong, L., Zhang, Y., Zhu, L., Wang, J., Wang, J., Du, Z. et al. 2018. Influence of isolated bacterial strains on the *in situ* biodegradation of endosulfan and the reduction of endosulfan- contaminated soil toxicity. Ecotoxicology and Environmental Safety 160(May): 75–83. doi:10.1016/j.ecoenv.2018.05.032.

Kumar, B. L. and Gopal, D. V. R. S. 2015. Effective role of indigenous microorganisms for sustainable environment. 3 Biotech. 5(6): 867–876. doi:10.1007/s13205-015-0293-6.

Kumar, N. S., Man, H. C. and Woo, H.-S. 2014. Biosorption of phenolic compounds from aqueous solutions using pine (*Pinus densiflora* Sieb) Bark powder. Bioresource Technology 9(3): 5155–5174. doi:10.15376/biores.9.3.5155-5174.

Kwon, B. G., Lim, H. J., Na, S. H., Choi, B. I., Shin, D. S. and Chung, S. Y. 2014. Biodegradation of perfluorooctanesulfonate (PFOS) as an emerging contaminant. Chemosphere 109: 221–225. doi:10.1016/j.chemosphere.2014.01.072.

Lacayo-Romero, M., Quillaguamán, J., Van Bavel, B. and Mattiasson, B. 2005. A toxaphene-degrading bacterium related to *Enterobacter cloacae*, strain D1 isolated from aged contaminated soil in Nicaragua. Systematic and Applied Microbiology 28(7): 632–639. doi:10.1016/j.syapm.2005.03.015.

Li, M. T., Hao, L. L., Sheng, L. X. and Xu, J. B. 2008. Identification and degradation characterization of hexachlorobutadiene degrading strain *Serratia marcescens* HL1. Bioresource Technology 99(15): 6878–6884. doi:10.1016/j.biortech.2008.01.048.

Li, P. S. and Tao, H. C. 2015. Cell surface engineering of microorganisms towards adsorption of heavy metals. Critical Reviews in Microbiology 41(2): 140–149. doi:10.3109/1040841X.2013.813898.

Liao, L., Chen, S., Peng, H., Yin, H., Ye, J., Liu, Z. et al. 2015. Biosorption and biodegradation of pyrene by *Brevibacillus brevis* and cellular responses to pyrene treatment. Ecotoxicology and Environmental Safety 115: 166–173. doi:10.1016/j.ecoenv.2015.02.015.

Lin, W. C., Chang-Chien, G. P., Kao, C. M., Newman, L., Wong, T. Y. and Liu, J. K. 2014. Biodegradation of polychlorinated dibenzo—dioxins by strain NSYSU. Journal of Environmental Quality 43(1): 349. doi:10.2134/jeq2013.06.0215.

Liu, M., Dong, F., Yan, X., Zeng, W., Hou, L. and Pang, X. 2010. Biosorption of uranium by *Saccharomyces cerevisiae* and surface interactions under culture conditions. Bioresource Technology 101(22): 8573–8580. doi:10.1016/j.biortech.2010.06.063.

Loss, O. E. M. and Yu, J. H. 2018. Bioremediation and microbial metabolism of benzo(a)pyrene. Molecular Microbiology 109(4): 433–444. doi:10.1111/mmi.14062.

Lu, M. 2012. Degradation of short chain polychlorinated paraffins by a new isolate: Tests in pure culture and sewage sludge. Journal of Chemical Technology & Biotechnology 88(7): 1273–1279. doi:10.1002/jctb.3971.

Maier, R. M. 2009. Bacterial growth. *In*: Maier, R. M., Pepper, I. L. and Gerba, C. P. [eds.]. Environmental Microbiology. Academic Press.

Martin, A. 1999. Biodegradation and Bioremediation. San Diego. Academic Press, USA.

Matsumoto, E., Kawanaka, Y., Yun, S. J. and Oyaizu, H. 2008. Isolation of dieldrin- and endrin-degrading bacteria using 1,2-epoxycyclohexane as a structural analog of both compounds. Applied Microbiology and Biotechnology 80(6): 1095–1103. doi:10.1007/s00253-008-1670-4.

Mejáre, M. and Bülow, L. 2001. Metal-binding proteins and peptides in bioremediation and phytoremediation of heavy metals. Trends in Biotechnology 19(2): 67–73. doi:10.1016/S0167-7799(00)01534-1.

Metcalf & Eddy, Inc., Tchobanoglous, G., Burton, F. and Stensel H. D. 2003. Wastewater Engineering: Treatment and Reuse. McGraw-Hill.

Michalak, I., Chojnacka, K. and Witek-Krowiak, A. 2013. State of the art for the biosorption process—A review. Applied Biochemistry and Biotechnology 170(6): 1389–1416. doi:10.1007/s12010-013-0269-0.

Mishra, A. and Malik, A. 2013. Recent advances in microbial metal bioaccumulation. Critical Reviews in Environmental Science and Technology 43(11): 1162–1222. doi:10.1080/10934529.2011.627044.

Mohseni, M., Akbari, S., Pajootan, E. and Mazaheri, F. 2019. Amine-terminated dendritic polymers as a multifunctional chelating agent for heavy metal ion removals. Environmental Science and Pollution Research doi:10.1007/s11356-019-04765-3.

Muralikrishna, I. V. and Manickam, V. 2017. Wastewater treatment technologies *In*: Environmental Management: Science and Engineering for Industry. Butterworth-Heinemann.

Navarro, R. R., Sumi, K. and Matsumura, M. 1998. Heavy metal sequestration properties of a new amine-type chelating adsorbent. Water Science and Technology 38: 195–201. doi:10.1016/S0273-1223(98)00528-9.

Navarro-Llorens, J. M., Tormo, A. and Martínez-García, E. 2010. Stationary phase in gram-negative bacteria. FEMS Microbiology Reviews 34(4): 476–495. doi:10.1111/j.1574-6976.2010.00213.x.

Nelson, L. N. 2006. Industrial Waste Treatment: Contemporary Practice and Vision for the Future. Butterworth-Heinemann, London.

Okeke, B. C., Siddique, T., Arbestain, M. C. and Frankenberger, W. T. 2002. Biodegradation of γ hexachlorocyclohexane (Lindane) and r-Hexachlorocyclohexane in water and a soil slurry by a Pandoraea species. Journal of Agricultural and Food Chemistry 50: 2548–2555.

Olafson, R. W., McCubbin, W. D. and Kay, C. M. 1998. Primary- and secondary-structure analysis of a unique prokaryotic metallothionein from a *Synechococcus* sp. cyanobacterium. Biochemical Journal 251: 691–699.

Orhon, D. and Artan, N. 1997. Modeling of Activated Sludge Systems. CRC Press. Basel, Switzerland.

Ozturk, S., Kaya, T., Aslim, B. and Tan, S. 2012. Removal and reduction of chromium by *Pseudomonas* spp. and their correlation to rhamnolipid production. Journal of Hazardous Materials 231-232: 64–69. doi:10.1016/j.jhazmat.2012.06.038.

Pal, T. K., Bhattacharyya, S. and Basumajumdar, A. 2010. Cellular distribution of bioaccumulated toxic heavy metals in *Aspergillus niger* and *Rhizopus arrhizus*. International Journal of Pharma and Bio Sciences 1(2): pp.BS57.

Pan, X., Lin, D., Zheng, Y., Zhang, Q., Yin, Y., Cai, L. et al. 2016. Biodegradation of DDT by *Stenotrophomonas* sp. DDT-1: characterization and genome functional analysis. Scientific Reports 6(October 2015): 1–10. doi:10.1038/srep21332.

Rahman, K. S. M., Rahman, T. J., Kourkoutas, Y., Petsas, I., Marchant, R. and Banat, I. M. 2003. Enhanced bioremediation of n-alkane in petroleum sludge using bacterial consortium amended with rhamnolipid and micronutrients. Bioresource Technology 90: 159–168.

Roberts, S. M., Turner, N. J., Tel, A. J. Willetts and Turner, M. K. 1995. Introduction to Biocatalysis Using Enzymes and Microorganisms. Cambridge University Press, Cambridge.

Rokade, R., Ravindran, S., Singh, P. and Suthar, J. K. 2018. Microbial biotransformation for the production of steroid medicament. *In*: Vijayakumar, R. and Raja, S. S. S. [eds.]. Secondary Metabolites—Sources and Applications, IntechOpen, DOI: 10.5772/intechopen.75149.

Rouf, A., Kanojia, V., Naik, H. R., Naseer, B. and Qadri, T. 2017. An overview of microbial cell culture. Journal of Pharmacognosy and Phytochemistry 6(6): 1923–1928.

Sakakibara, F., Takagi, K., Kataoka, R., Kiyota, H., Sato, Y. and Okada, S. 2011. Isolation and identification of dieldrin-degrading *Pseudonocardia* sp. strain KSF27 using a soil-charcoal perfusion method with aldrin trans-diol as a structural analog of dieldrin. Biochemical and Biophysical Research Communications 411(1): 76–81. doi:10.1016/j.bbrc.2011.06.096.

Shahi, A., Aydin, S., Ince, B. and Ince, O. 2016a. Evaluation of microbial population and functional genes during the bioremediation of petroleum-contaminated soil as an effective monitoring approach. Ecotoxicology and Environmental Safety 125(NOVEMBER): 153–160. doi:10.1016/j.ecoenv.2015.11.029.

Shahi, A., Aydin, S., Ince, B. and Ince, O. 2016b. Reconstruction of bacterial community structure and variation for enhanced petroleum hydrocarbons degradation through biostimulation of oil contaminated soil. Chemical Engineering Journal 306: 60–66. doi:10.1016/j.cej.2016.07.016.

Shahi, A., Ince, B., Aydin, S. and Ince, O. 2017. Assessment of the horizontal transfer of functional genes as a suitable approach for evaluation of the bioremediation potential of petroleum-contaminated sites: a mini-review. Applied Microbiology and Biotechnology doi:10.1007/s00253-017-8306-5.

Sherbert, G.V. 1978. The Biophysical Characterization of the Cell Surface. Academic press, London.

Singh, R. 2013. Microbial Waste Management. Edition 1. LAP LAMBERT Academic Publishing, ISBN: 3659377139.

Singh, M. S., Singh, S. and Singh, R. 2017. Microbial biotransformation: A process for chemical alterations. Journal of Bacteriology & Mycology: Open Access 4(2): 47–51. doi:10.15406/jbmoa.2017.04.00085.

Singh, R., Kumar, M., Mittal, A. and Mehta, P. K. 2016. Microbial enzymes: Industrial progress in 21st century. 3 Biotech. 6(2): 1–15. doi:10.1007/s13205-016-0485-8.

Smith, K. S. and Ingram-Smith, C. 2007. Genome analysis methanosaeta, the forgotten methanogen? Trends in Microbiology 15(4): 150–155. doi:10.1016/j.tim.2007.02.006.

Smitha, M. S., Singh, S. and Singh, R. 2017. Microbial biotransformation: A process for chemical alterations. Journal of Bacteriology & Mycology: Open Access 4(2): 47–51. doi:10.15406/jbmoa.2017.04.00085.

Speight, J. G. 2017. Environmental Inorganic Chemistry for Engineers. Butterworth-Heinemann, London.

Sperling, M.V. 2007. Basic Principles of Wastewater Treatment. IWA Publishing, London.

Sultan, A. and Reza, A. R. 2015. Steroids: A diverse class of secondary metabolites. Medicinal Chemistry (Los Angeles). 5(7). doi:10.4172/2161-0444.1000279.

Sun, S., Zhang, Z., Chen, Y. and Hu, Y. 2016. Biosorption and biodegradation of BDE-47 by *Pseudomonas stutzier*. International Biodeterioration and Biodegradation 108: 16–23. doi:10.1016/j.ibiod.2015.11.005.

Takagi, K., Iwasaki, A., Kamei, I., Satsuma, K., Yoshioka, Y. and Harada, N. 2009. Aerobic mineralization of hexachlorobenzene by newly isolated pentachloronitrobenzene-degrading *Nocardioides* sp. strain PD653. Applied and Environmental Microbiology 75(13): 4452–4458. doi:10.1128/AEM.02329-08.

Tezel, U., Tandukar, M. and Pavlostathis, S. G. 2011. Anaerobic biotreatment of municipal sewage sludge. *In*: Moo-Young, M. [ed.]. Comprehensive Biotechnology. Elsevier Science & Technology.

Ting, Y. F. and Praveena, S. M. 2017. Sources, mechanisms, and fate of steroid estrogens in wastewater treatment plants: a mini review. Environmental Monitoring and Assessment 189(4). doi:10.1007/s10661-017-5890-x.

Torres, M., Goldberg, J. and Jensen, T. E. 1998. Heavy metal uptake by polyphosphate bodies in living and killed cells of *Plectonema boryanum* (cyanophycae). Microbios. 96(385): 141–147.

Tsuji, N., Nishikori, S., Iwabe, O., Shiraki, K., Miyasaka, H., Takagi, M. et al. 2004. Characterization of phytochelatin synthase-like protein encoded by alr0975 from a prokaryote, *Nostoc* sp. PCC 7120. Biochemical and Biophysical Research Communications 315: 751–755.

Van Doesburg, W., Van Eekert, M. H. A., Middeldorp, P. J. M., Balk, M., Schraa, G. and Stams, A. J. M. 2005. Reductive dechlorination of β-hexachlorocyclohexane (β-HCH) by a Dehalobacter species in coculture with a *Sedimentibacter* sp. FEMS Microbiology Ecology 54(1): 87–95. doi:10.1016/j.femsec.2005.03.003.

Vijayaraghavan, K. and Yun, Y. S. 2008. Bacterial biosorbents and biosorption. Biotechnology Advances 26(3): 266–291. doi:10.1016/j.biotechadv.2008.02.002.

Villemur, R. 2013. The pentachlorophenol-dehalogenating *desulfitobacterium hafniense* strain PCP-1. Philosophical Transactions of the Royal Society of London. Series B, Biological Sciences 368(1616): 1–8. doi:10.1098/rstb.2012.0319.

Wang, L., Fan, D., Chen, W. and Terentjev, E. M. 2015. Bacterial growth, detachment and cell size control on polyethylene terephthalate surfaces. Scientific Reports 5: 1–11. doi:10.1038/srep15159.

Wang, X., Wang, X., Hui, K., Wei, W., Zhang, W., Miao, A. J. et al. 2018. Highly effective polyphosphate synthesis, phosphate removal, and concentration using engineered environmental bacteria based on a simple solo medium-copy plasmid strategy. Environmental Science & Technology 52(1): 214–222. doi:10.1021/acs.est.7b04532.

Water Environmental Federation. 2010. Nutrient Removal. WEF Manual of Practice No. 34. McGraw Hill Professional.

Wu, Z., Xie, M., Li, Y., Gao, G., Bartlam, M. and Wang, Y. 2018. Biodegradation of decabromodiphenyl ether (BDE 209) by a newly isolated bacterium from an e-waste recycling area. AMB Express. 8(1). doi:10.1186/s13568-018-0560-0.

Xing, J., Criddle, C. and Hickey, R. 1997. Effects of a long-term periodic substrate perturbation on an anaerobic community. Water Research 31(9): 2195–2204.

Xu, L., Chen, X., Li, H., Hu, F. and Liang, M. 2016. Characterization of the biosorption and biodegradation properties of *Ensifer adhaerens*: A potential agent to remove polychlorinated biphenyls from contaminated water. Journal of Hazardous Materials 302: 314–322. doi:10.1016/j.jhazmat.2015.09.066.

Yan, Y., Ma, M., Liu, X., Ma, W., Li, M. and Yan, L. 2017. Effect of biochar on anaerobic degradation of pentabromodiphenyl ether (BDE-99) by archaea during natural groundwater recharge with treated municipal wastewater. International Biodeterioration and Biodegradation 124: 119–127. doi:10.1016/j.ibiod.2017.04.019.

Yoshida, N., Morinaga, T. and Murooka, Y. 1993. Isolation and characterization of a heavy metal-binding protein from a heavy metal-resistant strain of *Thiobaccillus* sp. Journal of Fermentation and Bioengineering 76: 25–28.

Zambrano, M. M., Siegele, D. A., Almirón, M., Tormo, A. and Kolter, R. 1993. Microbial competition: *Escherichia coli* mutants that take over stationary phase cultures. Science 259(5102): 1757–1760.

Zhang, S., Xia, X., Xia, N., Wu, S., Gao, F. and Zhou, W. 2013. Identification and biodegradation efficiency of a newly isolated 2,20,4,40-tetrabromodiphenyl ether (BDE-47) aerobic degrading bacterial strain. International Biodeterioration and Biodegradation 76: 24e31.

CHAPTER 8

Energy

Duygu Nur Arabaci,[1] *Ilayda Dilara Unlu,*[1] *Aiyoub Shahi*[2]
and *Sevcan Aydin*[3],*

Introduction

Energy is a critical issue in the 21st century. With worldwide increasing demand of energy, in contrast to a depleting resource that is fossil fuels such as petroleum, natural gas and coal (IEA 2013), the need to find alternative ways to meet the energy demand of the technological world has become urgent. The search for a renewable energy source that is environmentally friendly, sustainable and economical has been recruiting an increasing number of researchers (Badar et al. 2018). At this rate of usage, it is estimated that fossil fuels, including oil, coal and natural gas, will be depleted within this century. Even still, the usage of fossil fuels results in increased global CO_2 emissions into the atmosphere, where it contributes to climate change as well as throwing ecosystems off balance. In addition, many studies have shown that microbial ecosystems react to concentration changes in CO_2 in the atmosphere as well as within the soil (Hungate et al. 1997, Norby et al. 2004), therefore causing complex environmental effects. Considering such issues, it is of utmost importance to develop novel ways of utilizing renewable sources for energy production for cleaner energy, termed bioenergy. Bioenergy consists products of various biological processes that utilize carbon. These biological processes may involve plant biomass, which is obtained from plants that absorb inorganic carbon, which is CO_2 in the air, through photosynthesis, and convert it into organic carbon, which is the type found within plants. The plant biomass that is obtained from such processes is utilized as feedstock for bioenergy applications. Still, the raw materials to be utilized as feedstock for bioenergy production are currently limited. Bioenergy production mainly utilizes feedstock from crops, which can include both food or feed crops as

[1] Department of Genetics and Bioengineering, İstanbul Bilgi University, Eyüpsultan, 34060 Istanbul, Turkey.
[2] Institute of Environment, University of Tabriz, Tabriz, Iran.
[3] Department of Genetics and Bioengineering, Nişantaşı University, Maslak, 34469, Istanbul, Turkey.
* Corresponding author: sevcan_aydn@hotmail.com

well as non-food crops; moreover, it can utilize waste or residues that are not actively used (Alam et al. 2012).

The biosphere contains its own sustenance, or as the common saying goes: nature provides. Alternative sources of energy include biofuels, solar energy, hydroelectric power generation and wind turbines. The biomass that can be obtained from nature has the potential to lead humanity into a closed energy economy and counter the problems that have arisen after the industrial revolution. While in percentage mass the biomasses may be low, they possess high levels of energy potential. Biomass can be utilized either raw or after being processed, to obtain energy. Unprocessed biomasses are mostly burnt as low value fuel. Or, they can be converted into biofuels to be utilized in various forms and for different purposes, such as to be utilized in transport, as in the case of biodiesel (Alam et al. 2012). Currently, the most applied and researched alternative to liquid fossil fuels is biofuels. Biofuels have been around since the beginning of humanity, in the form of wood, straw, charcoal and other unprocessed materials that have been commonly burned for heat. Biofuels can be liquid, solid or gaseous. Solid biofuels mostly consist of raw materials from plants, such as wood or solid agricultural wastes (Kumari and Singh 2018). Gaseous biofuels consist mainly of methane and hydrogen, obtained from the decay of biological materials (Kırtay 2011). Lastly, due to its ease in transport, processing and application, the most utilized biofuels are the liquid biofuels. This category consists mainly of bio-alcohols such as methanol and ethanol, oils and biodiesels. The modern biofuels however have been recently invented, and are obtained through complex chemical and biological processes, and these biofuels can be in forms such as biodiesel, bioethanol or synthetic gas.

Biofuels

Most of the energy generated from such renewable sources cannot currently be utilized by conventional transport engines that utilize diesel or gasoline, whereas several biofuel types do have compatibility with such engines. Bioethanol is the most commonly used commercial biofuel (Kumari and Singh 2018), and most of the cars in the US can use it as fuel without need for modification because of their ability to run on blends of E10, which includes 10% ethanol. However, different concentrations do require engine manipulations, which makes it harder to switch from conventional fossil fuels. Still, many countries have been making the effort: Brazil for example has made it mandatory to add ethanol into gasoline, and since 2007, the legal percentage of ethanol has become 25% (Matsuoka et al. 2010).

Biofuels utilize biomass from plants, microalgae or various sources of organic carbon. Recent developments in this area have been on utilizing agricultural feedstock which include corn (Farm-Energy 2019), soybean or palm oil (Transport Environment 2019), as well as from normally unused lignocellulosic biomass (Fatma et al. 2018) such as wood and plant wastes from paper and pulp industry, as well as from agriculture industry (Xie et al. 2016). This however may lead to new problems, such as eutrophication, resource depletion, reduced land space that could be used to meet food demands of the growing world population and deforestation of large areas, which has huge environmental impacts in the long run (Ward et al. 2014).

In bioethanol production, sugars are the main components utilized in the processes. Therefore, the high sugar content of microalgae species makes them potential targets for feedstock. With their high levels of glycogen and starch in their structures, some microalgae species that include Chlamydomonas, Chlorella, Dunaliella and Scenedesmus, as well as kelp have high potential when it comes to bioethanol production (Badar et al. 2018). While there are several ways of pretreatment for bioethanol production, the highest yield, about 84%, was obtained using steam explosion. On lignocellulosic biomass, using acids for pretreatment has been reported to be more efficient when used together with thermal pretreatment due to the conversion of the hemicellulosic components into sugars (Kumari and Singh 2018).

The valorization of the residues left over from bioethanol production in which algal biomass is used as feedstock has been an attractive research topic in the past few years. The second-generation bioethanol production was investigated for valorization of by-products and left-over residues for their potential of producing biogas. The biomass goes through several processes, the pretreatment and the enzymatic hydrolyzation, and the wastes that were left behind from these processes were utilized for biogas production (Rabelo et al. 2011).

During the bioethanol production using algae, algae oil is extracted, and even the leftover biomass can be valorized by being used as feedstock for biobutanol (Badar et al. 2018). As a replacement for fossil fuel derived products, biobutanol shows great promise thanks to the similarities it shares with them. Biobutanol that was obtained from algal biomass can be used as diesel fuel thanks to its chemically identical structure with butanol from fossil fuels and are therefore compatible with diesel engines (Kumari and Singh 2018). Furthermore, algal biomass and its residues left over from various processes can be utilized in other applications as well aside from bioenergy applications, as their rich composition of proteins, carbohydrates, vitamins and minerals also make them suitable candidates to be processed in an anaerobic digester in order to potentially be utilized as feedstock materials for animal and fishery feeds and as fertilizers (Badar et al. 2018).

Biogas

The processes to obtain energy from sources such as biomass and animal wastes require complex systems, but such processes occur naturally. The decay of such biomass and animal wastes are accomplished by decomposer organisms which include fungi and bacteria. These organisms break down the organic matter into its simpler components and return the nutrients and necessary chemicals back to the natural ecosystem. These processes usually take place in dark, warm and moist environments, with optimum activity in such conditions. The processes can be concluded via two types of bacteria, aerobic or anaerobic.

In aerobic digestion, the microorganisms and processes that take part require O_2, therefore these processes occur mostly above ground. Aside from aerobic bacteria, fungi and soil worms also take part in aerobic digestion in the presence of O_2. While during the decaying process some heat is released, this is not an efficient way of producing energy. The residue from aerobic digestion can be valorized as nutrient rich fertilizers.

Anaerobic digestion is the microbial process in which degradation of organic compounds occurs when there is no O_2 present in the environment. In such conditions, anaerobic bacteria oxidize the C in the biomass into CO_2 and reduces it to CH_4 (methane). The left-over residues from such processes are also rich in nutrients, such as N, and can be used as high value fertilizers. The CH_4 that is obtained through these processes is in gaseous form and mixed with CO_2. Biogas usually consists of mostly methane, and can be used as energy supply for cooking, heating and electricity generation, and the residues are utilized as fertilizers. The main goal of anaerobic digesters is mostly to optimize conditions for CH_4 production (Aydin 2017). Biomethane is an apt substitution for natural gas thanks to its identical traits (Moghaddam et al. 2019). Biomethane can be utilized for many applications such as fuel for transportation or for heat generation. Furthermore, while natural gas, which is methane that has been obtained through fossil fuels, is not sustainable as it depends on oil reserves that are being depleted, biomethane shows promise to reduce or even eradicate the global dependence on fossil fuels as a renewable, sustainable and green alternative since it is produced from organic matter, thus reducing the anthropogenic impact humans have on the atmosphere, and therefore leading to a decrease in concentration of methane and other greenhouse gases in the atmosphere.

Another biogas that is currently receiving much attention is the combination of biomethane and biohydrogen, termed Hythane®. Hythane is patented, and the patent holders' research that studied methane and hydrogen blends for usage in combustion engines reported promising results (Eden 2010). Moreover, the obtained methane can also be compressed and used as Bio-CNG (compressed natural gas), which provides an efficient and green alternative to gasoline or diesel fuels. However, the efficiency with which Bio-CNG burns is low, and therefore limits the economic and environmental potential:

It can only be burnt within a narrow range, burns slowly and at high temperatures (Bauer and Forest 2001). It has been reported that thanks to its higher mass specific heating value of hydrogen, the addition of even small amounts of it into bio-CNG widens the narrow range of flammability, and this mixing also reduces the burning time due to higher burning speed of hydrogen (Bauer and Forest 2001, Liu et al. 2012). Hydrogen is produced through dark fermentation most efficiently (Liu et al. 2012). This is because the production of hydrogen has many challenges, such as high cost of operating, the production of volatile fatty acids that have inhibitory effects, and the function of methanogens also within the systems. Therefore, the co-production of both gases in a two steps process offers a promising potential by raising the heat efficiency of bio-CNG, reducing the energy demand of the system and even increasing H_2 production overall. Both processes could complement each other's benefits as well as counter their disadvantages (Liu et al. 2012). Biohythane is currently replacing hythane as more research focuses on biological resources (Pasupuleti and Venkata Mohan 2015). Formally, biohythane can be defined as a mixture of 10–15% H_2, 50–55% CH_4 and 30–40% CO_2 and trace amounts of various other gases, all produced through two stage anaerobic digestion process. Anaerobic digestion offers an efficient way of utilizing VFAs for methane production, as there already exist methanogens (Hans and Kumar 2019).

Microalgae as feedstock

Renewable energy has received considerable attention in the last few decades as the fossil fuels are being depleted, and one of the major sustainable feedstock for biofuels, or sustainable transport fuels, has been micro-algae (Slade and Bauen 2013). In the recent years, the research has been redirected into utilizing microalgae in alternate ways of potential replacement of fossil fuels. Algae are a wide and diverse group of plant-like organisms that utilize photosynthesis. The species vary in size and shape, and they can be unicellular, such as Chlorella, or multicellular, such as giant kelp. Microalgae are utilized also in wastewater treatment as one of the most efficient biological treatment alternatives, which is termed phycoremediation. Their environmentally sustainable nature has allowed them to get much public support in the recent years when the urgent need to replace fossil fuels has become clear. Microalgae are considered as a valuable resource for biofuel production because of their ability to produce polysaccharides and triacylglycerydes, which are the raw materials utilized in biofuel production, in meaningful quantities (Slade and Bauen 2013). Algae are used in various production processes, which not only include liquid biofuels such as biodiesel but also extend to biogas, such as biomethane.

Several prospects of micro-algae as feedstock have propelled it above the alternatives, such as cost, sustainability, environmental impact and energy efficiency. While there are several methods of cultivation and species for the purpose of biofuel production, the cost of founding and maintaining a system mainly depends on the cost of CO_2, water and nutrients required for cultivation of microalgae, which can affect the total cost by over 50% (Slade and Bauen 2013). While the carbon dioxide emissions and energy consumption efficiency seem promising, the environmental impacts of such production need to be considered when designing a system, which puts constraints on system design and operations (Slade and Bauen 2013).

The microalgae can be cultivated all year round, providing an edge over crops. Even the highest yielding oilseed crops, such as 1190 l ha^{-1} for rapeseed or Canola (Schenk et al. 2008), 1892 l ha^{-1} for Jatropha (Singh and Gu 2010), and 2590 l ha^{-1} for Karanj (Pongamia pinnata) can't compete with the biodiesel yield of 58,700 l ha^{-1} of microalgae with 30% oil by weight. Next, the variety of microalgal species with oil content of 20–50% dry weight, and the adaptability to various systems proves beneficial (Kumari and Singh 2018). Also, the cultivation of algae species is faster, with exponential growth rates that are able to grow to double the biomass weight in 3.5 hours (Singh and Gu 2010).

One of such constraints is the water supply. For production of raw materials to be used for biofuel production using microalgae-based systems, the system would require a constant supply of fresh water. The fresh water may be required in cooling some photobioreactor designs, or as a constant addition to raceway pond systems to keep the water level stable and compensate for evaporation. However, microalgae systems still demand less fresh water than terrestrial crops (Dismukes et al. 2008). Another upside of using microalgae is that in the modern times where humanity is concerned about the diminishing fresh water sources (Richey et al. 2015), algae cultivation can utilize water that isn't otherwise used in many areas, such as saline sea water, or brackish aquifer water instead of fresh water (Singh and Gu 2010). However, unlike sea water, aquifer waters require pretreatment to be utilized efficiently due to their content of various chemicals that may inhibit the growth of algae, therefore increasing the energy required for cultivation process. Another alternative to reduce water consumption is using circulating water systems, which would also reduce nutrient loss. However, recirculation of the same water also increases the risk of contamination by microorganisms such as bacteria, fungi and viruses, as well as various substances that include organic and inorganic chemicals, remaining metabolites as well as the remains of dead algae cells. Raceway pond systems utilize large terrains for shallow ponds. This poses a problem, because the ponds require flat terrain, and the permeability and porosity of the soil affects pond sealing. The constraints for soil and topographic traits limit the land availability. Systems for algae cultivation also need to meet certain standards. The algae require nitrogen, phosphorus and potassium among various nutrients for growth, which have to be externally added into the system in most cases. Aside from these constraints, there's also a disadvantage that cannot be overlooked. It is common among algae strains to produce toxic chemicals within their natural life cycles, and these toxins range from ammonia to polypeptides and polysaccharides that affect the normal functioning of living organisms. Examples for such effects include acute ones such as paralytic shellfish poison, which causes severe reactions in many organisms, even leading to death on certain occasions and there are effects that cause damage through long term exposure, such as carrageenan toxins produced in red tides, which lead to tissue changes that may result in cancerous growths or ulcer. These toxins may leak into the environment, and operating such systems may become cumbersome due to such effects (Singh and Gu 2010).

There are also environmental issues to consider when it comes to micro-algae cultivation for biofuel production on a large enough scale to overthrow fossil fuels from mainstream usage. The cultivation of microalgae could result in various changes in the environment depending on the location of the cultivation system. The effects are not necessarily negative; depending on the system, it can even lead to increased benefits for the cultivation environment. There is not enough data to conclude the effects of specific systems, and especially genetically modified strains on the ecological systems. The primary impact of such a micro-algae cultivation system on the ecosystem it is located in essentially lies in water management, the input that is taken from the environment and consumed during the process, as well as the effluent from the cultivation system, its contents as well as the volume and where it goes. Therefore, strict monitoring of the ecosystem of the environment around the microalgae cultivation systems plays a crucial role in the future of such facilities (Singh and Gu 2010).

Another requirement is a constant and reliable source of carbon dioxide. Algae are assumed to have 50% carbon mass, which means that the production of dry algal biomass of 1 kg demands 1.82 kg of CO_2. In application, the demand for CO_2 becomes even higher than this due to external losses (Chisti 2007). In raceway ponds, the factors of CO_2 loss rate to consider include pond depth, mixing velocity, pH, friction coefficient of the lining and alkalinity. These factors decrease the CO_2 utilizing efficiency in theoretical equations, which can be between 20% and 90%. In applications, the numbers fall even further; in open raceway pond systems, the CO_2 fixation efficiencies have been observed to fall below 10%, which averages at 35% in closed thin layer cultivation systems. A higher CO_2 fixation efficiency of approximately 75% is also reported in tubular photobioreactor systems. However, the demand and utilization of CO_2 by microalgae also means that microalgae cultivation can improve air quality and help with the increasing concentrations of CO_2 in the air.

Another input to consider for algae cultivation is energy consumption. Cultivation of algae requires specific conditions, especially temperature, therefore the systems may need temperature control in order to maintain high levels of productivity, especially in photobioreactors. The need for cooling and heating may increase the energy demand of the systems. The algae cultivation systems use electricity during operations, as well as natural gas in order to dry the produced algae, which translates into fossil fuel consumption that comes with many negative aspects, therefore reducing the benefits of the green sustainable biofuels.

Lignocellulosic biomass

Ethanol has taken the spotlight in recent years, being considered the biofuel from sustainable resources that has the highest potential to replace the usage of fossil fuels in commercial use. This has led to much effort being spent on optimization and development of wide-spread commercial production of bioethanol. While lignocellulosic material from non-food crops is being utilized as feedstock in second generation biofuel production (Alam et al. 2012), various issues that revolve around utilizing crops and are therefore competing with land space for food crops have been decreased.

While there are many bagasse that can be utilized for biofuel production, the most abundant lignocellulosic biomass, especially in tropical countries is sugarcane, especially cultivated in Brazil. Two Hundred and Eighty kg of bagasse can be obtained from 1 tons of the sugarcane cultivated in Brazil, while the harvest rates for sugarcane were estimated to be 568.50 million tons from the 2011/2012 harvest season (UNICA 2011). Only 50% of the residues from these sugarcanes are utilized for energy source in distillery plants (Rabelo et al. 2011).

Aside from obtaining raw materials from crops that are directly raised for it, or the non-used parts of food crops, industrial wastes can also be utilized. Industrial wastes usually contain high amounts of lignocellulosic biomasses, especially pulp and paper industry. Therefore, the developing of a concept of a biorefinery that can treat and valorize such waste and producing fuels and chemicals to be fed back into the economy has been receiving great attention and offers many benefits both economically and environmentally.

Such a biorefinery would be utilizing much of the raw material rather than only taking what is necessary and the rest going to waste. Maximizing the usage potential of all parts of the biomass and its intermediates and optimizing processes according to market situations and biomass availability offer significant advantages (Rabelo et al. 2011).

The lignocellulosic material is made up of three main components: cellulose, hemicellulose and lignin; however, efficient degradation of lignin remains a challenge due to its recalcitrant backbone, and most biofuels utilize cellulose as their main component. To overcome this challenge, four main steps have been developed for the second-generation biofuel production, which consist of pretreatment to degrade lignin and expose cellulose for usage, hydrolysis via the utilization of enzymes or acid catalysts to release the monomeric sugars, fermentation for the released sugars to be converted into ethanol and distillation to obtain the purified final product (Xie et al. 2016).

A pretreatment method is hydrogen peroxide. It is commonly used in pulp and paper industry as well as various industries that utilize cellulose, where it is used as bleaching agent. The product obtained from such pretreatment is also preferable as hydrogen peroxide degrades into water and oxygen, leaving no residue behind on the final product. Another advantage of this method is that there are little to no secondary product formation.

The pentoses produced by microorganisms are also a potential resource for ethanol production; however, the usage of microorganisms as feedstock in current processes in ethanol production has led to disappointing results with low yields. While the existing microorganisms have not been promising, their genetically modified versions might, and the genetically engineered microorganisms that can process pentoses into ethanol have been a promising area of research. Even still, there are no current established methods that can be utilized in industrial applications.

Anaerobic digesters for food waste as feedstock

The economy in the world is currently not a closed economy, which means that much of the world's wastes are not utilized, and much of the components depend on consumption and waste; however, this is not a sustainable system. Anaerobic digestion is a process that is carried out by anaerobic microorganisms in which various organic materials and biomasses from biodegradable wastes, obtained from various wastes from industries such as food and paper and pulp industry, are processed, through which biogas is obtained, and it plays an important part in recovering valuable materials (Aydin 2017). An advantage of anaerobic digestion processes for biogas production is that the residue left behind during the process also contains valuable nutrients, and it can be used in agricultural applications (Xu et al. 2018). Anaerobic digestion applications are preferred over other methods of obtaining bioenergy because they can be utilized on a wide range of substrates, and do not require the removal of impurities or moisture (Fagerström et al. 2018).

Anaerobic digestion has four stages: hydrolysis, acidogenesis, acetogenesis and methanogenesis (Xie et al. 2016). Hydrolysis step consists of breaking down of complex organic matters, such as carbohydrates, proteins and lipids into sugars, amino acids and fatty acids, respectively. In the next stage, acidogenesis, the resulting monomers of the previous stage are converted mainly into VFAs, such as acetate, propionate, and butyrate (Aydin 2017). Acetogenesis is the next stage where VFAs are converted into acetic acid, H_2, CO_2 by acetogenic bacteria. In the final stage of methanogenesis, the resulting products of acetic acid, H_2 and CO_2 are converted into methane and CO_2 (Hans and Kumar 2019). Even though anaerobic digestion methods are compatible for various types of organic wastes, it is most commonly used to treat impure and low-quality wastes that are not otherwise efficiently valorized.

Another advantage that the anaerobic digestion systems have is the flexible conditions, as the process can be carried out in both small- and large-scale digesters, and the location of such a system is not a constraint. Anaerobic digesters are currently applied for various purposes such as treatment of wastewater, sewage and animal manure (Aydin 2017). However, as the production of biogas from waste gains even more attention, these treatments are constantly developing. A recent addition to the promising substrates to be utilized for energy production through anaerobic digesters is food waste (Paritosh et al. 2017), due to their high energy content, large quantity, and wide availability. 1.3 billion tons of food, which accounts for one third of the total food produced worldwide, is wasted and for 30 years ahead that number is expected to increase to 3.4 billion tones (The World Bank 2018). Recovering energy from waste therefore does not only encompass wastewater, sewage and industrial waste, but also extends into food wastes that are discarded, and the recycling of the energy discarded in such wastes remains a challenge and a requirement for a sustainable society.

Food wastes can be pre-consumer, which includes waste during production and processing such as the non-utilized parts of fruits and vegetables, the roots and stems as well as husks, peels, pomace and various other residues that are produced during the processing of such foods. These wastes contain starch, protein, sugars, lipids and various other valuable compounds that are either returned to the agricultural economy by being processed into animal feed or utilized in industrial applications for extraction or synthesizing of high demand chemicals in pharmaceutical or cosmetic industries, as well as in food industry (Xu et al. 2018). Current treatment facilities that process food wastes do so together with animal manure or sewage sludge, and this co-digestion provides necessary conditions and nutrients required for the anaerobic digestion process for both. However, anaerobic digestion methods still face various challenges, such as high costs of transportation and operation, and technical challenges such as volatile fatty acid accumulation and process instability, and low buffer capacity also remain (Aydin 2017). Still, thanks to the rapid developments in molecular sciences, the microbial community structures have been available for analysis, allowing for construction of more complex systems for biogas and organic material production (Aydin et al. 2015a, b, c).

For food waste treatment, different anaerobic digestion processes with optimal conditions for each substrate have to be designed specifically for each food waste type. The highest reported methane potential that can be achieved for each type of food waste has been higher than other potential substrates such as sewage and agricultural waste such as animal manure and lignocellulosic biomass, with a range of between 0.3–1.1 m³ CH_4/kg VS-added. The highest level of methane yield achieved with 1.1 m³ CH_4/kg-VS-added has been reported as the waste that is inclusive of fats, oils and grease, which is fitting with the calculated methane potential of lipids (1.014 m³/kg VS), whereas carbohydrates score much lower; an example of this is glucose with a methane potential of 0.37 m³/kg VS. Aside from substrate specific optimization requirements, the systems for anaerobic digestion also need tight control during process as well, since it is possible for harmful intermediate chemicals to be accumulated during the process should the treatment process for anaerobic digestion be not properly optimized or controlled, leading to instability of the process and lower yield efficiency of methane.

Conclusion

From the screening and selection of the strains with desired traits that has been in effect longer, to the more recent cultivation of modified strains of improved algae, molecular genetic techniques have been invaluable for the improvement of algae cultivation systems. In recent years, supplying energy has been a growing concern and with the rapid developments in technology, the utilization of biological resources in the effort for a sustainable energy economy is becoming easier. The alternative to depleting fossil fuels come in the form of biofuels, fuels that derive their feedstock from renewable substances, such as lignocellulosic biomass, algae and wastes that are not valorized. While the earlier biofuels have focused on lignocellulosic biomass from plants, or wastes, the more recent studies have been focusing on microorganisms as biomass, mainly microalgae, which draws interest thanks to its flexibility of many strains and easy and efficient cultivation, all year-round production and efficient yields, as well as environmental benefits. As a new and promising area of research, systems that utilize microalgae have still a long way to go before becoming commonplace; however, their potential cannot be underestimated.

References

Alam, F., Date, A., Rasjidin, R., Mobin, S., Moria, H. and Baqui, A. 2012. Biofuel from algae—is it a viable alternative? Procedia Engineering 49: 221–227. doi: 10.1016/j.proeng.2012.10.131.

Aydin, S., Ince, B., Cetecioglu, Z., Arikan, O., Ozbayram, E. G., Shahi, A. et al. 2015a. Combined effect of erythromycin, tetracycline and sulfamethoxazole on performance of anaerobic sequencing batch reactors. Bioresource Technology 186: 207–214. https://doi.org/10.1016/j.biortech.2015.03.043.

Aydin, S., Ince, B. and Ince, O. 2015b. Application of real-time PCR to determination of combined effect of antibiotics on Bacteria, Methanogenic Archaea, Archaea in anaerobic sequencing batch reactors. Water Research 76: 88–98. doi: 10.1016/j.watres.2015.02.043.

Aydin, S., Shahi, A., Ozbayram, E. G., Ince, B. and Ince, O. 2015c. Use of PCR-DGGE based molecular methods to assessment of microbial diversity during anaerobic treatment of antibiotic combinations. Bioresource Technology 192: 735–740. doi:10.1016/j.biortech.2015.05.086.

Aydin, S. 2017. Anaerobic digestion. Waste Biomass Management—A Holistic Approach 1: 14. doi: 10.1007/978-3-319-49595-8_1.

Aydin, S., Yıldırım, E., Ince, O. and Ince, B. 2017. Rumen anaerobic fungi create new opportunities for enhanced methane production from microalgae biomass. Algal Research 23: 150–160. doi: 10.1016/j.algal.2016.12.016.

Badar, S. N., Mohammad, M., Emdadi, Z. and Yaakob, Z. 2018. Algae and their growth requirements for bioenergy: A review. Biofuels 1–19. doi:10.1080/17597269.2018.1472978.

Bauer, C. G. and Forest, T. W. 2001. Effect of hydrogen addition on the performance of methane-fueled vehicles. Part I: effect on S.I. engine performance. International Journal of Hydrogen Energy 26: 55–70. doi:10.1016/s0360-3199(00)00067-7.

Chisti, Y. 2007. Biodiesel from microalgae. Biotechnology Advances 25(3): 294–306. doi:10.1016/j.biotechadv.2007.02.001.

Dismukes, G., Carrieri, D., Bennette, N., Ananyev, G. and Posewitz, M. 2008. Aquatic phototrophs: efficient alternatives to land-based crops for biofuels. Current Opinion in Biotechnology 19(3): 235–240. doi: 10.1016/j.copbio.2008.05.007.

Eden. 2010. Eden Annual Report. Available from: <http://edeninnovations.com/>

Fagerström, A., Seadi, T., Rasi, S. and Briseid, T. 2018, August. The Role of Anaerobic Digestion and Biogas in the Circular ... Retrieved from https://www.ieabioenergy.com/wp-content/uploads/2018/08/anaerobic-digestion_web_END.pdf.

Farm-Energy. 2019, April 3. Corn for Biofuel Production. Retrieved from https://farm-energy.extension.org/corn-for-biofuel-production/.

Fatma, S., Hameed, A., Noman, M., Ahmed, T., Shahid, M., Tariq, M. et al. 2018. Lignocellulosic biomass: A sustainable bioenergy source for the future. Protein and Peptide Letters 25(2): 148–163. https://doi.org/10.2174/0929866525666180122144504.

Hans, M. and Kumar, S. 2019. Biohythane production in two-stage anaerobic digestion system. International Journal of Hydrogen Energy 44(32): 17363–17380. doi:10.1016/j.ijhydene.2018.10.022.

Hungate, B. A., Holland, E. A., Jackson, R. B., Chapin, F. S., Mooney, H. A. and Field, C. B. 1997. The fate of carbon in grasslands under carbon dioxide enrichment. Nature 388(6642): 576–579. https://doi.org/10.1038/41550.

IEA, World energy outlook 2013, International Energy Agency, Paris, France (2013).

Kırtay, E. 2011. Recent advances in production of hydrogen from biomass. Energy Conversion and Management 52(4): 1778–1789. doi: 10.1016/j.enconman.2010.11.010.

Kumari, D. and Singh, R. 2018. Pretreatment of lignocellulosic wastes for biofuel production: A critical review. Renewable and Sustainable Energy Reviews 90: 877–891. doi: 10.1016/j.rser.2018.03.111.

Liu, Z., Zhang, C., Lu, Y., Wu, X., Wang, L., Wang, L. et al. 2012. States and challenges for high-value bio-hythane production from waste biomass by dark fermentation technology. Bioresour. Technol. 135: 292–303. https://doi.org/10.1016/j.biortech.2012.10.027.

Matsuoka, S., Ferro, J. and Arruda, P. 2009. The Brazilian experience of sugarcane ethanol industry. *In Vitro* Cellular and Developmental Biology Plant 45(3): 372–381. https://doi.org/10.1007/s11627-009-9220-z.

Moghaddam, E. A., Ericsson, N., Hansson, P. and Nordberg, Å. 2019. Exploring the potential for biomethane production by willow pyrolysis using life cycle assessment methodology. Energy, Sustainability and Society 9(1). doi:10.1186/s13705-019-0189-0.

Norby, R. J., Ledford, J., Reilly, C. D., Miller, N. E. and O'Neill, E. G. 2004. Fine-root production dominates response of a deciduous forest to atmospheric CO_2 enrichment. Proceedings of the National Academy of Sciences of the United States of America 101(26): 9689–9693. https://doi.org/10.1073/pnas.0403491101.

Paritosh, K., Kushwaha, S. K., Yadav, M., Pareek, N., Chawade, A. and Vivekanand, V. 2017. Food waste to energy: an overview of sustainable approaches for food waste management and nutrient recycling. BioMed. Research International 1–19. doi:10.1155/2017/2370927.

Pasupuleti, S. and Venkata Mohan, S. 2015. Single-stage fermentation process for high-value biohythane production with the treatment of distillery spent-wash. Bioresource Technology 189: 177–185. doi: 10.1016/j.biortech.2015.03.128.

Rabelo, S., Carrere, H., Filho, R. M. and Costa, A. 2011. Production of bioethanol, methane and heat from sugarcane bagasse in a biorefinery concept. Bioresource Technology 102(17): 7887–7895. doi:10.1016/j.biortech.2011.05.081.

Richey, A., Thomas, B., Lo, M., Reager, J., Famiglietti, J., Voss, K. et al. 2015. Quantifying renewable groundwater stress with GRACE. Water Resources Research 51(7): 5217–5238. doi: 10.1002/2015wr017349.

Schenk, P., Thomas-Hall, S., Stephens, E., Marx, U., Mussgnug, J., Posten, C., et al. 2008. Second generation biofuels: high-efficiency microalgae for biodiesel production. BioEnergy Research 1(1): 20–43.

Singh, J. and Gu, S. 2010. Commercialization potential of microalgae for biofuels production. Renewable and Sustainable Energy Reviews 14(9): 2596–2610. doi: 10.1016/j.rser.2010.06.014.

Slade, R. and Bauen, A. 2013. Micro-algae cultivation for biofuels: Cost, energy balance, environmental impacts and future prospects. Biomass and Bioenergy 53: 29–38. doi:10.1016/j.biombioe.2012.12.019.

The World Bank. 2018, September 20. Global Waste to Grow by 70 Percent by 2050 Unless Urgent Action is Taken: World Bank Report. Retrieved from https://www.worldbank.org/en/news/press-release/2018/09/20/global-waste-to-grow-by-70-percent-by-2050-unless-urgent-action-is-taken-world-bank-report.

Transportenvironment. 2019, January 24. Palm Oil and Soy Oil for Biofuels Linked to High Rates Of Deforestation. Retrieved from https://www.transportenvironment.org/press/palm-oil-and-soy-oil-biofuels-linked-high-rates-deforestation-new-study.

Ward, A., Lewis, D. and Green, F. 2014. Anaerobic digestion of algae biomass: A review. Algal Research 5: 204–214. doi: 10.1016/j.algal.2014.02.001.

Xie, S., Ragauskas, A. and Yuan, J. 2016. Lignin conversion: Opportunities and challenges for the integrated biorefinery. Industrial Biotechnology 12(3): 161–167. doi: 10.1089/ind.2016.0007.

Xu, F., Li, Y., Ge, X., Yang, L. and Li, Y. 2018. Anaerobic digestion of food waste challenges and opportunities. Bioresource Technology 247: 1047–1058. doi:10.1016/j.biortech.2017.09.020.

CHAPTER 9

Next Generation Sequencing
Advances and Applications in the World of Bacterial Diversity

T. J. Sushmitha, Meora Rajeev and
*Shunmugiah Karutha Pandian**

Introduction

Bacteria are the most diverse and abundant group in Earth's biota. Bacterial diversity encompasses a huge spectrum of bacterial species that are crucial for an ecosystem. New discoveries suggest that bacterial biota extends deep into the Earth's crust and thrive in sub-surfaces of ocean and land. It is estimated that the total carbon biomass of subsurface microorganisms is equal to all marine and terrestrial plants (Pedersen 2000). Bacteria being omnipresent in nature forms the foundation of the biosphere and they are crucial players of the environment as they exclusively catalyze processes nourishing all life on Earth. These prokaryotes masterpiece a suitable function in dynamics, shaping the trophic web networks and in the remineralisation of organic matter and thus are the engine that regulates biogeochemical cycle (Feng et al. 2009). They also act as the key indicators for any change in environmental and nutrient parameters.

Diversity study has become a major concern these days as our understanding on role of bacterial community (collection of diverse bacteria coexisting in an environment at a given time) is increasing day by day. An age of culture-dependent studies gave insights into bacterial physiology and genetics but the study of whole community composition was a hard task because of its small size and enormous diversity (Moughan et al. 2012). Diversity assessments for natural bacterial community compositions was conventionally influenced by cultivable species, but results obtained from the use of advance molecular methods proposed that culture-dependent techniques has led to underestimation of bacterial diversity.

Department of Biotechnology, Alagappa University, Science Campus, Karaikudi 630003, Tamil Nadu, India.
* Corresponding author: sk_pandian@rediffmail.com

Analysis of bacterial DNA and RNA revealed bacterial diversity far greater than predicted. The recent advances in molecular fingerprinting techniques based on Polymerase Chain Reaction (PCR), viz., Denaturant Gradient Gel Electrophoresis (DGGE), Terminal-Restriction Fragment Length Polymorphism (T-RFLP), Amplified Ribosomal DNA Restriction Analysis (ARDRA), and Automated Approach for Ribosomal Intergenic Spacer (ARISA) has paved the way to discover many new species and enlarged the view of bacterial diversity. These approaches failed to expose least abundant species and was a tiresome process. These flaws led to the development of Next-Generation Sequencing (NGS) that dominates the study of bacterial diversity in the current era.

Transition from genome to metagenome *(Sequencing to Next-Generation Sequencing)*

Starting with the discovery of DNA's molecular structure by James Watson and Francis Crick (Watson and Crick 1953), boundless paces have been made to comprehend the complexity and diversity of genomes in environment. **Genome**, the entire set of genes or the whole genetic material present in a cell, has been analyzed thoroughly to identify and to describe the full potential of particular cell. Though bacteria are omnipresent and crucial for functioning of an ecosystem, bacterial genomics and diversity were restricted to an individual bacterial cell. The information attained from the genome sequence of many bacterial individuals was utilized for our understanding of microbial physiology, genetics and evolution (Moughan et al. 2012, Allen and Banfield 2005). This approach was limited to only 0.1 to 1% of entire bacterial community where *in vitro* cultivation is cumbersome for plenty of bacteria in an environment and few remain uncultivable. There started the isolation of **Metagenome**—genome of entire bacterial community from any habitat and **Metagenomics**—named by Handelsman as the study of environmental or whole community genome (Handelsman 2004).

Metagenome or the community DNA takes advantage over genomic DNA because of the following: (i) it involves many fragments of DNA from different individuals with number of fragments indicating bacteria's abundance, (ii) it is directly isolated from environmental samples such as soil (Yang et al. 2017), marine sediment (Song et al. 2012, Aravindraja et al. 2013), drinking water (Hou et al. 2018), river water (Ma et al. 2016) and marine biofilm (Rajeev et al. 2019), which helps in study of its impact on its environment, and (iii) it focuses on existing community from an environment will potentially support ecological studies.

A brief history of DNA sequencing

DNA sequencing is a technique to determine the precise order of nucleotides within a DNA molecule. DNA sequencing has revolutionized the knowledge of genome and has enabled the development and improvement of metagenomics especially in understanding bacterial diversity. A brief timeline of major milestones achieved in DNA sequencing is depicted (Fig. 1). Initially, two methods: (i) Chain Termination by Frederick Sanger (Sanger et al. 1977) and (ii) Chemical degradation by Maxam

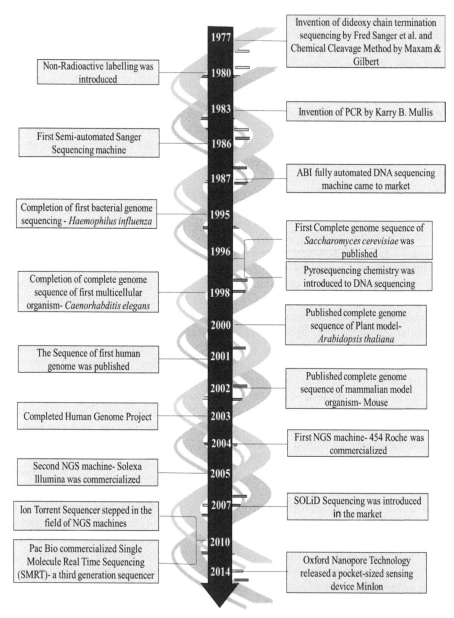

Fig. 1. Milestones in DNA sequencing. The image shows the timeline representing the major breakthroughs in the history of DNA sequencing.

and Gilbert (Maxam and Gilbert 1977) were developed. Sequencing by radiolabelled methods underwent various improvements and many attempts were made to automate the sequencing process. The development of fluorescence labelling, capillary gel electrophoresis and automated sequencer (Hunkapiller et al. 1991) by Leroy Hood in 1986 (ABI automated DNA sequencer) increased the daily throughput of sequencing data and the rate of complete genome sequencing of various organisms followed. Sanger sequencing—a gold standard technique (Besser et al. 2018)—ruled the era of genomics due to its associated simplicity and made monumental accomplishments including completion of sequencing entire human genome. This method was the prime preference from 1977 to 2005 and is still ideal for culture-dependent bacterial identification and routine molecular assays.

Though the Sanger's method completed various genome projects such as first bacterial genome—*Haemophilus influenza* in 1995 (Fleischmann et al. 1995), *Caenorhabditis elegans* in 1998, *Arabidopsis thaliana* in 2000 and Human Genome (Venter et al. 2001), complications faced during these projects demanded a faster and cheaper sequencing technologies that led to the development of Second-Generation Sequencing (SGS), also referred as NGS. The onset of NGS created a huge impact in sequencing age and made the understanding of complex biological organisms easier. NGS has the command to replace and supersede DNA microarrays as NGS was able to sequence and quantify all genes without any prior information of a particular sequence. The capability of NGS to sequence whole genome of many interrelated organisms through massively parallel sequencing at significantly low cost has established a strong foundation in comparative and evolutionary studies of bacteria in an unimaginable way. Simultaneously, various platforms with different chemistry flourished the sequencing market.

While NGS technologies are enormously powerful, the drawbacks such as very short read length and use of PCR for template amplifications reduced the efficiency of NGS. A new generation of sequencing—Third-Generation Sequencing (TGS)—technology targeting a single DNA molecule was improved, such that no synchronization is required (a drawback in NGS). This generation of sequencing has the potential to exploit high processivity of DNA polymerase and to overcome biases of PCR and dephasing. TGS technology also has a shorter time to generate the outcome. Comparison of all three generation DNA sequencing technology is given (Table 1). Therefore, while the sequencing technologies have revolutionized the study of genome, there is a need to exploit these technologies to understand and unlock the potential of metagenomics in bacterial diversity.

First-generation sequencing (FGS)

Initial and preliminary efforts to sequence DNA were challenging and cumbersome. Cohesive ends of bacteriophage lambda with a size of 12 bases was the first to be sequenced by Wu using primer extension method (Wu and Kaiser 1968). Gilbert and colleagues in the year 1973 transcribed lactose-repressor binding site with 24 bases into RNA and sequenced those fragments (Gilbert and Maxam 1973).

DNA sequencing was totally decoded and made accessible in 1977 by two renewed methods—Chemical cleavage method of Maxam and Gilbert (Maxam and Gilbert 1977) and dideoxy Chain-termination method of Sanger (Sanger et al. 1977). Though the method by Maxam and Gilbert was popular and arose with simple principles, it had whole lot difficulties. The disadvantage lies in the use of huge amount of radioactive materials and chemicals, e.g., hydrazine, which is a neurotoxin. Difficulty is also encountered with procedures involving radioactive

Table 1. Comparison of sequencing generations.

	First generation	Second generation	Third generation
Sequencing strategy	Shotgun sequencing	Massively Parallel Sequencing	Single Molecule Sequencing (SMS)
Sequencing approach	Sequencing by Termination (SBT)	Sequencing by Synthesis and Ligation (SBS & SBL)	Single Molecule Real-Time (SMRT) sequencing
Read length	~ 500–1000 bp	~ 200–400 bp	> 1000 bp
Read accuracy	High	High	Low
Sequencing throughput	Low	High	High
Cost	High cost per base, Low cost per run	Low cost per base, High cost per run	Low cost per base, High cost per run
Sample preparation	PCR amplification/cloning	PCR amplification followed by library construction	PCR amplification not required, single template sequencing
Duration	~ 1 day	~ 2–3 days	> 1 day
Data generated	Least complex	Most complex	Complex
Results	Base calls with quality values	Base calls with quality values	Base calls with quality values
Platforms	ABI Prism	454/Roche, Ion Torrent, Illumina, SOLiD	-SMAT- Pac Bio -Oxford Nanopore's MinION

Adapted from Schadt et al. 2010. A window into third-generation sequencing. Hum. Mol. Genet. 19(R2): R227–R240.

labelling, X-ray film development and cleavage reactions that needed exact ratio of chemicals. This method confirmed about 200–300 bases every few days. Sanger sequencing superseded Maxam and Gilbert's method due to its simplicity and ease use of chemicals.

There was a major breakthrough in DNA sequencing with Sanger's Chain termination method. This technique relies on enzymatic synthesis of DNA *in vitro* that uses (i) single strand DNA as template (ii) primers for extension (iii) Klenow fragment of DNA polymerase-I that lacks proof reading activity (iv) ordinary deoxyribonucleotides (dNTPs) as building blocks (v) chemically modified and radiolabeled deoxyribonucleotides called di-deoxynucleotides (ddNTPs) that lacks 3'-OH group required for the formation of phospho-diester bond and therefore unable to form a bond with next dNTP. The first genomes sequenced by the Sanger sequencing are ΦX174 (genome size of 5374 bp) and bacteriophage λ (with length of 48,501 bp).

Sanger sequencing was further improved in following years, which utilizes fluorescent labeled chain terminator ddNTPs (Dye-terminator sequencing) that allows reaction to take place in single tube instead of four (Smith et al. 1985) and use of capillary based electrophoresis for sensitive detection (Swerdlow and Gesteland 1990). Both of these advances contributed in the development of automated Sanger sequencer with the capacity to sequence with a read-length of nearly one kilo base (kb). Applied biosystems is the first company that commercialized Sanger sequencing machine called ABI Prism 370 (Kchouk et al. 2017).

In the automated sequencing method, template DNA can be prepared in one of two following ways: first, target DNA fragment can be amplified by PCR using primers that flank the target. Second, shotgun *de novo* sequencing to analyze long DNA fragment. In shotgun sequencing, a long strand of DNA molecule is fragmented into short segments, cloned into high-copy-number plasmid and used to transform *Escherichia coli*. This allowed commercialization of DNA sequencing to sequence complex genomes. Shotgun sequencing is the latest of first-generation sequencing that analyzes longer fragments by cloning of overlapping DNA sequences and assembling them into one contiguous sequence (a contig) *in silico*. The development of NGS also took place concurrently (Fig. 2).

The second/next-generation sequencing technologies

Accomplishment of Human Genome Project (HGP) stimulated a way to develop "massively parallel sequencing" or "next-generation sequencing" that succeeded all electrophoretic sequencing methods. Advent of new generation of sequencing technology broke the limitation of FGS and are different in terms of high-throughput, cost and massively parallel analysis. The fundamental advantages of NGS, relative to FGS include the following: (i) instead of hundreds, millions of reactions can be generated in parallel (ii) cloning of DNA fragments is not required as preparation of libraries is carried out in cell-free system and (iii) bases can be detected without electrophoresis.

The term "second/next-generation sequencing" is used in reference to various commercialized platforms, viz., Roche/454 Genome Sequencer (Roche Diagnostics

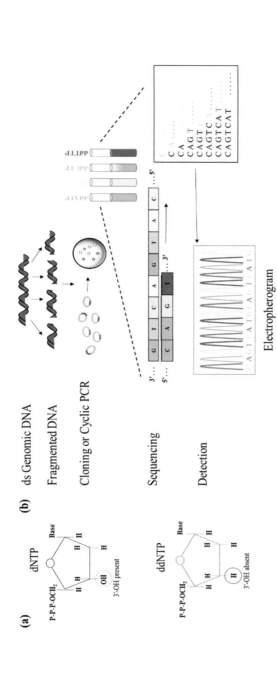

Fig. 2. Sanger's chain-termination sequencing method. (a) Diagrammatic representation of deoxyribonucleotide triphosphate (dNTP) and di-deoxyribonucleotide triphosphate (ddNTP). Naturally occurring deoxyribose sugar has 3′-OH group that helps in formation of phospho-diester bond with next nucleotide. The absence of 3′-OH group inhibits the DNA nucleotide extension. This chemistry was used by Sanger to synthesize di-deoxyribose sugar that lacks 3′-OH moiety. (b) Work flow of shotgun Sanger sequencing. The genomic DNA is fragmented and cloned into plasmids. Sequencing is performed in a micro-scale volume, where normal dNTPs (in all four tubes) in ample quantities and ddNTPs (individual base in four different tubes) in limited quantities are added along with other sequencing reagents (manual sequencing). In automated Sanger sequencing through cyclic PCR, four bases of ddNTP are fluorescently labelled with different colors (shown as green for ddATP, yellow for ddGTP, blue for ddCTP and red for ddTTP) which aids in their detection. Sequencing by DNA polymerase is stopped when a ddNTP is incorporated. This creates DNA fragments of various sizes. The sequenced products are run on a high resolution capillary gel electrophoresis (automated Sanger sequencing) and bases are detected and represented as an Electropherogram. The color of peaks in electropherogram represents the fluorescence color used for four base. Electropherogram is the raw data for Sanger sequencing that represents the quality of the sequence
(Image reproduced from Shendure et al. 2017. DNA sequencing at 40: past, present and future. Nature 550(7676): 345).

Corp., Branford, CT, USA), Illumina (Illumina Inc., San Diego, CA, USA), AB SOLiD System (Life Technologies Corp., Carlsbad, CA, USA) and Ion Torrent (Life Technologies, South San Francisco, CA, USA) that utilizes cyclic-array sequencing. The comparison of different NGS platforms along with their sequencing chemistry and various features such as run times, read length and yield is given (Table 2). Sequencing approaches utilized by all commercial NGS platforms fall into two wide categories: Sequencing by Synthesis and Sequencing by Ligation.

Sequencing by synthesis (SBS): This approach involves detection of upcoming nucleotide by synthesizing and incorporating nucleotides complimentary to template strand by DNA polymerases (polymerases dependent approach). As a single nucleotide is incorporated, the signal gets detected by means of fluorophore attached to the nucleotide or change in ionic concentration of membrane when a nucleotide enters. Cyclic Reversible Terminator Technology (CRTT) by Illumina platform and Single Nucleotide Addition (SNA) by Roche 454 and Ion Torrent are the two chemistry involving SBS approach.

Sequencing by ligation (SBL): SBL approach (Ligase dependent approach) uses multiple bases at a time to sequence. A partially degenerated probe (octamer or nonamer oligonucleotide) encoding one or two known bases is fluorescently labelled and allowed hybridization with existing template. DNA ligase ligates the oligonucleotide to pre-existing probe for detection and imaging. Fluorescent labels are then cleaved to regenerate 5′-end for next set of oligonucleotide ligation. AB SOLiD System utilizes SBL approach for sequencing.

All NGS platforms are PCR-based technologies that rely on amplification of template strands to numerous folds. These sequencing technologies comprise number of steps that are clustered broadly as (i) template/library preparation, (ii) sequencing, (iii) imaging and (iv) data analysis. The diverse chemistry behind the utilization of these protocols and their unique combinations distinguishes one platform from other and defines the sort of data generated from each platform. Even though these platforms have diverse sequencing biochemistry and array generation, their workflow remains conceptually similar. Template preparation and library construction—the preliminary step involves clonal *in vitro* amplification of template strand that creates dense multiplexing with millions of template strands immobilized on a substrate. Simple approaches, termed "emulsion PCR" and "bridge amplification", generates polonies (PCR colonies with many copies of particular library fragment) that are used to create library. Recent advances include clonal "nanoballs" generation with rolling circle amplification (RCA) (Drmanac et al. 2010). These polonies act as centers for SBL/SBS reaction with its own individual DNA template that facilitates massively parallel DNA sequencing. Sequencing platforms accumulate information generated from millions of SBL/SBS reactions in chorus, thus enabling the sequencing of many millions of DNA templates in parallel.

Clonal in vitro amplification

NGS uses cell-free system to amplify DNA templates that overcomes the arbitrary losses of genomic DNA sequences—an advantage over bacterial cloning used in

Table 2. Comparison of NGS platforms.

	454 Roche	Illumina	Ion Torrent	SOLiD
Amplification	Emulsion PCR	Bridge Amplification	Emulsion PCR	Emulsion PCR
Chemistry	Sequencing by Synthesis (SBS) Pyrosequencing	Sequencing by Synthesis (SBS) Reversible Terminator	Sequencing by Synthesis (SBS) Proton Detection	Sequencing by Ligation (SBL)
Yield (Gb/run)	0.7	1–60	1	3
Highest average read length	700 bp	300 bp (overlapping paired end sequencing)	400 bp	75 bp (paired end sequencing)
Error rate	1	< 0.1	1	< 0.06
Advantages	Long read	High Throughput low cost	Least expensive	Low error rate due to two base error coding system
Disadvantage	Expensive reagents and high error rates in homopolymer sequences	Short reads and long run time	Homopolymer errors	Long runtime

Adapted from De Mandal et al. 2015. Microbial ecology in the era of next generation sequencing. Next Generat. Sequenc. & Applic. S: 1–2.

FGS. Chief techniques include emulsion PCR and bridge amplification that generate clonally clustered template library for massively parallel DNA sequencing. Once after the library is prepared by pair-end or mate-pair target adapters, they undergo emulsion PCR or bridge amplification.

Emulsion PCR (emPCR)

It is a bead based PCR method that allows mass parallelization of sequencing reactions by enormously increasing the content of DNA that can be sequenced in a solitary run. Emulsion PCR consists of a library of random DNA fragments with mate-pair targets that is attached to a bead in an aqueous droplet (micelle). The bead's surface encompasses primers with universal priming sites having oligonucleotide complimentary to the target DNA ends, thus allowing each bead to associate with single DNA fragment. These beads containing complex genomes then undergo water-in-oil emulsion PCR with common PCR reagents where each bead is coated with millions of clonal DNA population (Fig. 3). After the successful PCR amplification and emPCR bead enrichment, millions of these beads are immobilized either in polyacrylamide gel (Polonator) or deposited into Pico titer plate as one bead per well. The 454 Roche, the Polonator and SOLiD platforms rely on emPCR for clonal DNA amplification.

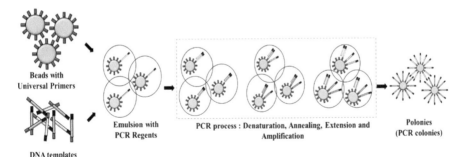

Fig. 3. Schematic representation of Emulsion PCR. The 454 Roche, Ion Torrent and SOLiD platforms rely on emulsion PCR (emPCR) for clonal amplification. Ideally, emPCR functions as a micro-PCR reactor in a bulk oil phase. emPCR contains a reaction mixture with *in vitro* constructed adapter flanked DNA templates (shown as black and red adapters flanking templates) and numerous droplets of PCR reaction mixture in oil phase. Each droplet bead is enriched with 5' attached adapter sequence (shown as red color on oil droplet) complimentary to adapters flanking templates. At low template concentration, DNA template is complemented with the adapters in each bead, resulting in one/zero template per bead. The emulsion undergoes PCR process: Denaturation, Annealing and Extension where polonies of clonal clusters are generated. Each bead is enriched with millions of copies from single DNA template. These encapsulated bead-DNA complex is diluted to compartmentalize one bead per well in a Picotiter plate to perform the sequencing (Image reproduced from Metzker. 2010. Sequencing technologies—the next generation. Nature Review Genetics 11(1): 31).

Bridge amplification (Cluster PCR)

This technique involves a solid phase amplification of *in vitro* built adaptor flanked library that avoids the use of emulsions and favors the multiplication directly on a slide. In this approach, both forward and reverse primers are densely coated and tethered covalently on a solid substrate. The primers are attached at their 5' ends by a flexible linker randomly, such that amplification products rising from any of the single DNA template molecule persist to remain immobilized locally and clusters to a point of origin on an array. These primers offer complementary sequences to ssDNA templates for the attachment and amplification. Consequently, a subsequent bridge amplification occurs that produces clonally amplified DNA populations from every individual DNA template (Fig. 4). Resulting clusters comprise ~ 1000 clonal amplicons of a single template (Shendure and Hanley 2008) and several millions of clustered clonal amplicons can be generated at distinguishable locations that are sequenced in parallel in a single run. The ratio of both primers to the DNA templates on the solid substrate determines the surface density of clonally amplified clusters. Hence, detailed control over the concentration of template and primers eases the amplification of DNA templates and maximizes the cluster density in a non-overlapping and localized way to maintain the spatial integrity and to avoid overcrowding. This approach has been called as bridge amplification, owing to replicating and extending DNA strands that arcs over (creating bridge) to lead the subsequent round of DNA extension through adjoining surface bound primer. Illumina platforms rely on bridge amplification for sequencing.

The benefits of NGS at present counterbalance several disadvantages. The most noticeable of these comprises read length and accuracy of raw base calling. Read lengths are currently far shorter than conventional Sanger sequencing and accurate base calls generated on average by all new NGS platforms are ~ tenfold less accurate than base calls produced by conventional dideoxy chain-termination approach

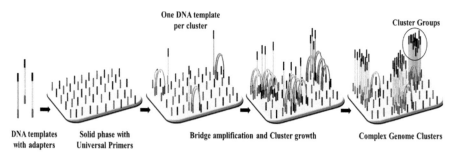

Fig. 4. Depiction of Bridge amplification. The Illumina platform relies on bridge amplification (cluster PCR) for sequencing. Here, forward and reverse sequences that complement the template adapters are adhered covalently to the solid flow-cell surface. They are named as P5 and P7 adapters (shown as black and red respectively, on solid surface). The adapter flanked DNA templates (2 adapters are used that flank both ends of the DNA) are allowed to anneal with P7 adapters. The adapters attached through 5' end remain flexible allowing other side of the DNA template to flip and form bridge complementing with adjacent located P5 adapters. The hybridized template DNA-adapter is extended by DNA polymerases to produce a double stranded bridge. Double stranded bridge is denatured to form two single stranded DNA sequences. These two sequences act as template for next round of PCR cycle and the process continues to generate multiple copies of same parent DNA template forming clusters. Each cluster generates thousand copies from one original sequence. Finally, all the reverse strands attached to P7 adapters are cleaved and washed away leaving a cluster only with forward strands (P5 adapter adhered). 3'-OH group is blocked to inhibit unwanted DNA priming and SBS based sequencing is performed with the forward strands (Image reproduced from Shendure and Hanley. 2008. Next-generation DNA sequencing. Nature Biotechnology 26(10): 1135).

(Shendure and Hanley 2008). Improvements are being made with respect to these parameters to increase the efficiency and precision of sequencing.

Major NGS platforms

454/Roche System

Roche 454 system was the first successful commercial available NGS platform in 2004. This platform uses SBS approach that works on the principle of "pyrosequencing". It is a non-electrophoretic and bioluminescence approach that measures the inorganic pyrophosphate released on nucleotide incorporation by DNA polymerase and by converting it to visible light from the cleavage of an Oxyluciferin through a series of enzymatic reactions (Fig. 5). The intensity of light produced is directly proportional to the number of nucleotides incorporated by DNA polymerase. Contrasting to other sequencing chemistry that relies on modified nucleotides to culminate DNA sequencing, pyrosequencing approach deploys the DNA polymerase by addition of any single dNTP in a limited quantity that is incorporated to complement the dNTP of template. After extension of the nucleotide, DNA polymerases is halted and reinitiated for the next set of dNTP release.

In brief, the protocol of pyrosequencing includes the pre-incubation of amplicon-bearing beads with DNA polymerase (e.g., *Bacillus stearothermophilus*— Bst polymerase) and proteins as single-stranded binding protein that are dropped onto a picotiter well plate in the way that each well accommodates one bead. Smaller beads containing immobilized enzymes (ATP sulfurylase and luciferase), adenosine 5-phosphosulphate (APS) and luciferin are also added. During each cycle, one species of dNTP (unlabeled) is introduced into each well and the incorporation event of dNTP on templates is detected by Charge Coupled Device (CCD). This renders array based massive parallel sequencing using pyrosequencing reaction that immensely increases the sequencing throughput. The order of the sequence and its respective intensity of light is recorded as "flowgrams" that disclose the underlying DNA sequence (Fig. 5).

The key benefit of 454/Roche platform is its read length. 454 FLX instrument produces about ~ 4,00,000 reads per solitary run with the read length of 200–300 bp. The foremost limitation and dominant error of this platform is insertion-deletion, than substitution that is allied to homopolymer sequences (consecutive repeats of same base, viz., TTTT or GGG or AAA). Signal intensity is the only way to detect the length of the homopolymer sequence, as no terminating group is available to prevent consecutive incorporations of bases at a particular cycle promoting more error rate than the discernment of incorporation of bases as compared to non-incorporation.

Pros. The main advantages of this platform include long reads (1 kb maximum) and relatively fast run time (~ 23 h). The long read provides easier map to reference genome for metagenomic applications.

Cons. This platform provides relatively low throughput (~ 1 million reads and 700 mb sequence data) with high reagent cost. The error rates are high in case of homopolymer repeats.

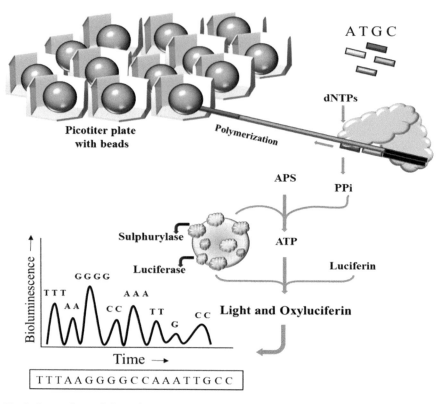

Fig. 5. Sequencing work flow of 454 Roche. Clonally amplified beads from emulsion PCR is sequenced through SBS approach. Compartmentalized beads in Picotiter well plate serves as a matrix for storage of DNA polymerase, Sulfurylase (shown as orange cloud), Luciferase (shown as green cloud), apyrase (not shown) and dNTPs. Pyrosequencing in Roche consists of series of downstream enzymatic reaction that result in production of light when a nucleotide is incorporated. Firstly, cycles of four dNTPs are added separately (either A/T/G/C). At each nucleotide incorporation, template is extended and releases PPi, whose quantity is equal to the number of nucleotides incorporated. Adenosine 5′-phosphosulfate (APS) aids in conversion of PPi to Adenosine Tri-Phosphate (ATP) quantitatively in the presence of ATP sulfurylase. The generated ATP converts the luciferin to Oxyluciferin (produces visible light) by involving the enzyme luciferase. The intensity of the light produced is usually detected at 550 nm by a charge coupled device (CCD). Apyrase is nucleotide degrading enzyme that degrades all non-incorporated nucleotides in the reaction mixture. A time interval usually of 65 seconds is maintained to allow incorporation of known nucleotides added and degradation of non-incorporated nucleotide. The light generated is represented as 'pyrogram' (similar to electropherogram of Sanger di-deoxy sequencing). Next cycle is repeated with next known set of nucleotide (Image reproduced from Mardis, E.R. 2008. The impact of next-generation sequencing technology on genetics. Trends in Genetics 24(3):133–141).

Illumina/Solexa genome analyzer

Illumina platform was initially developed in 2006 and its advances over intervening years tremendously gained huge output, quality and much reduction in cost, making it a dominant high-throughput sequencer in NGS market. In Illumina sequencer, a DNA molecule is sequenced using SBS approach and cyclic reversible termination

(CRT) method. CRT employs reversible terminators and a special DNA polymerase to incorporate modified dNTPs into the growing chain. A reversible terminator has 3'-end blocked using blocking groups such as 3'-O-allyl-2' and 3'-O-azidomethyl and a fluorophore is added to a base.

The workflow of Library preparation follows same steps as discussed above, viz., fragmentation of DNA molecules, enzymatic trimming, adenylating and ligation of adapter sequences. An adequate concentration of library fragments is placed on flow cell (composed of glass and eight microfluidic channels, each containing covalently bound adapter sequences complementary to the library adapter) to generate millions of clusters by using bridge amplification. In next step, primers bind to DNA fragments to facilitate the sequencing. Cyclic reversible termination tactic of sequencing comprises three subsequent steps (i) nucleotide incorporation (ii) fluorescence imaging and (iii) cleavage of blocking group to facilitate next cycle. All four nucleotides are added at each cycle as their fluorescence is uniquely detected for each individual nucleotide. Sequencing occurs by incorporating only single nucleotide in each cycle as a blocking group attached at 3'-OH position in reversible terminator, precluding the addition of next nucleotide within a cycle. After incorporation of single nucleotide by polymerase, the excess of unincorporated nucleotides are washed away and blocking group as well as fluorescent moiety is removed by chemical treatment. Imaging is performed to determine the incorporated nucleotide. The terminator dye utilizes four discriminable flour colors to distinguish between four different nucleotides and the sequence of the template is detected by reading the colors when DNA polymerase adds successive nucleotides (Fig. 6).

Illumina can also function by paired-end sequencing, i.e., it can sequence both ends of a fragment of defined size. The Illumina instrument is designed to read the other end of the generated cluster by initially removing the synthesized DNA strands through denaturation followed by regeneration of opposite end clusters by performing a limited bridge PCR that makes a second read with improved signal-to-noise ratio. The reverse primer is primed after the amplification by releasing the opposite end fragments from the flow cell using a cleaving reagent and the reverse strand is sequenced same as forward. During sequence analysis, a base calling algorithm designed for the Illumina platform aligns the sequences and eliminates the poor quality reads by maintaining Phred quality score (≥ 20) followed by other quality assessment.

Pros: At present, Illumina dominates the NGS industry. This platform has compatibility with most of the library preparation protocols. Moreover, Illumina offers higher sequencing throughput as compared to all other platforms with least per-base cost. An average read length of 300 bp is companionable for all applications.

Cons: The platform particularly faces the problem of sample loading and sequence complexity. Samples with low-complexity, viz., 16S rRNA gene library needs dilution or mixed with PhiX reference library to generate diversity. Limitations also embrace dephasing (improperly extended bases that either fall one base behind or after), which reduces the signal-to-noise ratio that decreases the sequencing efficiency.

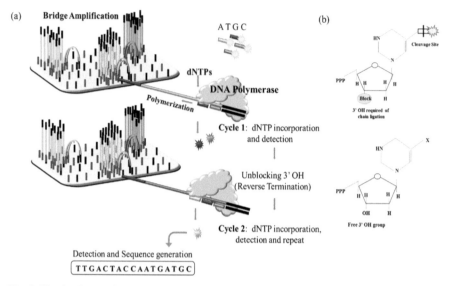

Fig. 6. Illumina Sequencing. a) Library preparation using bridge amplification and sequencing through reversible termination method. The nucleotides used here are modified with a chemical group blocking the 3'-OH site and are labelled with four different fluorescence color. A sequencing primer is introduced to the forward sequence of the cluster and is hybridized to the annealing site of adapter sequence. The fluorescent labelled reversible terminator nucleotides are allowed one by one to get polymerized. On each modified nucleotide incorporation, the DNA polymerase halts due to absence of 3'-OH group. This halt aids in detection of fluorescence through CCD. Once the fluorescence is detected, the next cycle begins by adding the chemical that de-blocks the 3'-OH group and by washing off all the non-incorporated nucleotides. b) shows the fluorescence cleavage in nitrogenous base and 3'-OH blockage site. After the incorporation of a nucleotide, 3'-OH is unblocked and fluorescence is cleaved (Image reproduced from Escalante et al. 2014. The study of biodiversity in the era of massive sequencing. Revista Mexicana de Biodiversidad 85(4): 1249–1264).

Ion Torrent/Semiconductor sequencing

A semiconductor sequencing technology was commercialized in 2010 by Ion Torrent, a company that was later adapted by Life Technologies. In contrast to other NGS platforms, this platform hires entirely different sequencing chemistry. The detection of incorporated nucleotide is not based on imaging of fluorescence signal but on the detection of hydrogen ions (H^+) released as a by-product of nucleotide incorporation during polymerization.

The workflow of this sequencer is analogous to 454 Roche. Adapter flanked template amplicon library is constructed and amplified in emulsion PCR. The beads in emulsion are transferred to a microchip plate (the workhorse of Ion Torrent Sequencer) that permits only one bead per well and the clonally amplified template population in each bead is read in parallel. Ion Torrent applies SBS approach, wherein synthesis of a strand by nucleotide incorporation releases hydrogen ion (H^+) that changes the pH of the solution. The complementary metal-oxide semiconductor (CMOS) technology detects the signal of H^+ and is translated to voltage that is deciphered as sequence read in computer software pipelines. The voltage intensity is

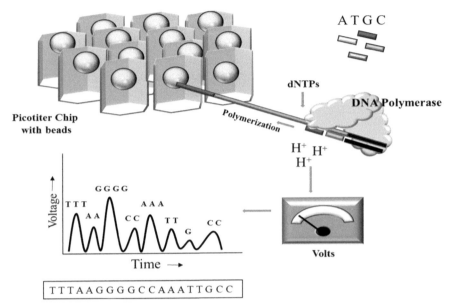

Fig. 7. Hydrogen ion based sequencing workflow of Ion Torrent platform. Beads from the emulsion PCR are compartmentalized into individual wells of the Picotiter well plate (one bead per well). SBS approach is employed to determine the nucleotide of the template DNA sequence. Known nucleotides are sprayed in sequence and are incorporated by DNA polymerase, which releases one H^+ ion per incorporation. The released H^+ ion creates a positive charge resulting in the change in current and pH. The net change in current is detected by a special device called pH-sensitive field effect transistor (pHFET). The non-incorporated nucleotides are washed and the next cycle is repeated (Image reproduced from Escalante et al. 2014. The study of biodiversity in the era of massive sequencing. Revista Mexicana de Biodiversidad 85(4): 1249–1264).

directly proportional to the number of nucleotides integrated to the template strand (Fig. 7). In this sequencing platform, individual dNTP are added as there is no detectable difference for H^+ released from dNTPs (A, T, G, and C).

Pros: Since there is lack in utilization of any recorder such as CCD, or detector for fluorescence, the advantage over time period has drastically increased in Ion Torrent sequencer facilitating completion of single run within 4 h.

Cons: Similar to 454 Roche, this platform also faces the disadvantage of insertion-deletion error while reading homopolymer sequences. Uninterrupted incorporation of bases from homopolymer sequences may result in misjudging the signals causing either excess of bases (insertion) or shortage of a base (deletion). Secondly, as there is no measurable differences in H^+ for all four bases (A, T, G, C), addition of individual dNTPs is required to all consecutive cycles. Any slipup can cause substitution error in parallel.

ABI/SOLiD^{TM} sequencer

Sequencing/Supported by Oligo Ligation Detection (SOLiD) is the first sequencing platform that was acquired by Applied Biosystems in 2006 and uses multiple bases to sequence. This platform exploits SBL with hybridization approach by employing partial degenerative primers with one or two known bases.

In brief, the protocol involves the preparation of pair/mate-pair end adapter flanked DNA template library that is clonally amplified with emulsion PCR, equivalent to 454 Roche and Ion Torrent. Paramagnetic beads carrying clonally amplified sequences contribute platform to massively sequence the DNA template by ligation-hybridization process. These beads are disorderly immobilized on a solid planar flowcell surface. The instrument uses an octamer oligonucleotide primer with ligation site at first two/three bases (universal priming site), cleavage site at fifth/sixth bases and fluorescence labelling at the eighth base. Four colored fluorescence labelling codes the 16 possible two base combinations (two-base encoding system). Once, eight base oligonucleotide probe integrates to the adapter sequence, the sequencing proceeds by the use of DNA ligase that ligates at $5'-PO_4$ group. Incorporated bases of the hybridized probe are identified through fluorescence detection. Cleavage occurs at the fifth position of the oligonucleotide probe along with the universal bases leaving free $5'-PO_4$ moiety for the next probe to ligate. Succeeding round of octamer probe ligation occurs at every fifth base position (e.g., 5, 10, 15, and 20). After every ligation and detection, denaturation follows for next ligation cycle at n-1 position, i.e., one position behind the first ligated base (Fig. 8). This permits each base to read twice allowing much lesser error rate than other sequencing platforms and thereby the method gives the sequencing data with least error rate compared to other platforms.

Pros: SOLiD is claimed to have the least error rate (~ 99.94% accuracy) as all the bases are read twice. In addition, next to Illumina, SOLiD provides high sequencing throughput.

Cons: It is limited to sequence only a short read length (75nt maximum) with longest run time compared to other platforms. Disadvantage also lies in its two base color encoding system, as misjudging may lead to misinterpretation of the succeeding bases.

Next-generation sequencing technologies to assess microbial community

Individual organisms cannot survive in any natural environments, rather they live in a dynamic consortium that conserves the ecosystem and its functionality. Insight into the dynamics of organism's community is the principal tactic for better understanding of functionality of an ecosystem and this can be made possible by extracting the total genomic DNA (metagenome) of the consortium directly from its natural environment. Environmental samples such as soil, seawater and marine sediment contain huge genetic diversity encompassing a range of microorganisms from Eukarya, Archaea and Bacterial domain.

Advent of massively parallel sequencing has dramatically revolutionized the field of microbial diversity and allows the generation of thousand to million reads of nucleotide sequences in single instrument run at a relatively low cost. The capability of these technologies to sequence deeper and deeper allows us to investigate bacterial

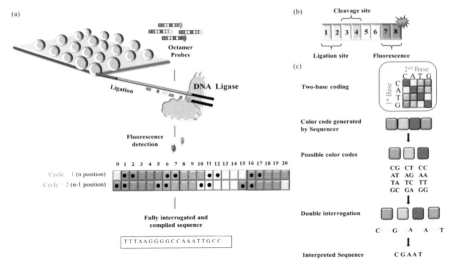

Fig. 8. Schematic illustration of SOLiD Sequencing. (a) Similar to 454 Roche and Ion torrent, DNA fragments for SOLiD sequencing is amplified through emulsion PCR and the magnetic beads are deposited onto a flow-cell glass slide and the beads are linked through covalent bonds. SOLiD sequencing relies on 5' DNA ligation using DNA ligase enzyme. Firstly, a universal primer is complemented with the adapter region of the DNA template and ligation mediated sequencing is performed with fluorescently labelled 8-mer dual probes (as only first two bases are incorporated). The first two bases of the 8-mer is ligated (nth position) and the rest are cleaved along with fluorescent detection. The next primer binds to the next bases and this usually continues till 35 bp (reads can also be extended). Once one set of primer is ligated, the dsDNA formed is denatured and the second set is annealed at n-1 position. This allows the base to be detected twice, which increases the accuracy of the sequencer. The cycle is repeated till the whole DNA template is read. (b) Octamer used in SOLiD sequencing. The octamer depicts the ligation site at first two bases (universal priming site), cleavage site at fifth base and fluorescence labelling site at the eighth base. (c) Reading a two base coding system. Each fluorescent color codes for two bases and they are decoded by single base deletion and aligning the rest of the sequence (Image reproduced from Mardis, E. R. 2008. Next-generation DNA sequencing methods. Annual Review of Genomics and Human Genetics 9: 387–402).

diversity at various taxonomic levels. Visions of inaccessible organisms are now enlarged with the field of metagenomics. Any NGS study follows some common steps such as environmental sample collection, extraction of high quality metagenomics DNA, library construction, sequencing, data preprocessing and diversity analysis. Major studies carried out at various environments on bacterial diversity are enlisted in Table 3.

With no uncertainty, NGS ruled the utilization of metagenome than any other culture-independent techniques and has provided unprecedented access to metagenome for inferring the bacterial diversity at a comparative higher depth. NGS has addressed solutions for many interrogations including classical population genetics of microbial community ecology (McCormack et al. 2013) and molecular systematic by providing base to assess natural microbial phenomenon comprising of biogeochemical cycles, evolutionary events, identification of novel genes, global microbial interactions (Escalante et al. 2014) and bacterial community dynamics at temporal and spatial scales. Two kind of NGS strategies are utilized in bacterial

Table 3. Bacterial diversity studies carried out with different NGS platforms and strategies.

Purpose	Platform	Strategy	References
Environmental studies			
Microbial diversity in deep sea water masses of deep North Atlantic and its hydrothermal vents	454 Roche	Amplicon sequencing	Sogin et al. 2006
Bacterial diversity analysis in marine sediment of Palk Bay, India	Illumina	Amplicon sequencing	Aravindraja et al. 2013
Comparative analysis of bacterial community composition in seawater and marine biofilms	Illumina	Amplicon sequencing	Rajeev et al. 2019
Bacterial diversity in the atmosphere of Mexico City	Ion Torrent	Amplicon sequencing	Serrano et al. 2018
Gene profiling of planktonic and biofilm microbial population	Illumina	Amplicon sequencing	Celikkol et al. 2016
Cross-biome metagenomic analyses of soil microbial communities from cold deserts, hot deserts, forests, grasslands, and tundra	Illumina	Amplicon and shotgun sequencing	Fierer et al. 2012
Temperature driven changes in benthic bacterial diversity	Illumina	Amplicon sequencing	Hicks et al. 2018
Microbial ecology of Thailand tsunami and non-tsunami affected areas	454 Roche	Amplicon sequencing	Somboonna et al. 2014
Comparison of the active and resident community of a coastal microbial mat	Illumina	Amplicon sequencing	Cardoso et al. 2017
Food microbiology			
Bacterial diversity of meat and seafood spoilage microbiota	Illumina	Amplicon sequencing	Poirier et al. 2018
Bacterial diversity of Cocoa bean fermentation	454 Roche	Shotgun sequencing	Illeghems et al. 2012
Diversity of Danish raw milk and cheese	454 Roche	Amplicon sequencing	Masoud et al. 2012
Microbial diversity of Brazilian kefir grains	454 Roche	Amplicon sequencing	Leite et al. 2012
Anthropogenic activities			
Benthic bacterial diversity in coastal areas contaminated by heavy metals, (PAHs) and (PCBs)	Illumina	Amplicon sequencing	Quero et al. 2015
Anthropogenic drivers of bacterial community changes in the Estuary of Bilbao and its tributaries	Illumina	Amplicon sequencing	Aguirre et al. 2017
Elevated nutrients change bacterial community composition of young marine biofilms	454 Roche	Amplicon sequencing	Lawes et al. 2016

Table 3 contd. ...

... Table 3 contd.

Purpose	Platform	Strategy	References
Microbiome of built cultural heritage	Ion Torrent	Amplicon sequencing	Adamiak et al. 2018
Bacterioplankton dynamics within a large anthropogenically impacted Urban Estuary	454 Roche HiSeq Illumina	Amplicon and shotgun sequencing	Jeffries et al. 2016
Microbiome diversity			
Diversity of the healthy human microbiome	454 Roche	Shotgun sequencing	Huttenhower et al. 2012
Microbiome profile of fungivorous thrips *Hoplothrips carpathicus* at different developmental stages	Illumina	Amplicon sequencing	Kaczmarczyk et al. 2018
Effect of different diets on microbiome diversity and fatty acid composition of rumen liquor in dairy goat	Illumina	Amplicon sequencing	Cremonesi et al. 2018
Cat Flea microbiomes	Illumina	Amplicon sequencing	Vasconcelos et al. 2018
Core microbiome in Cattle Rumen	Ion Torrent	Amplicon sequencing	Wirth et al. 2018
Clinical microbiology			
Cervical intraepithelial neoplasia disease progression is associated with increased vaginal microbiome diversity	Illumina	Amplicon sequencing	Mitra et al. 2015
Vaginal microbiomes in reproductive-age Women with Vulvovaginal Candidiasis	Illumina	Amplicon sequencing	Liu et al. 2013
Diversity of deep dentinal caries lesions in teeth with symptomatic irreversible pulpitis	Illumina	Amplicon sequencing	Rôças et al. 2016
Detection of *Listeria monocytogenes* in CSF from Three Patients with Meningoencephalitis	Ion Torrent	Shotgun	Yao et al. 2016
Asthma-associated differences in microbial composition of induced sputum	454 Roche	Amplicon sequencing	Marri et al. 2013
NGS comparison (within platforms, hypervariable regions of 16S rRNA gene, Amplicon Vs. Shotgun sequencing) in bacterial diversity			
16S rRNA gene amplicon and shotgun sequencing comparison in temporal microbial community	Illumina	Amplicon and shotgun sequencing	Poretsky et al. 2014
16S rRNA gene amplicon and shotgun sequencing comparison in microbiome	Illumina, Ion Torrent	Amplicon and shotgun sequencing	Clooney et al. 2016

Table 3 contd. ...

... Table 3 contd.

Purpose	Platform	Strategy	References
16S rRNA gene amplicon and shotgun sequencing comparison in microbiome	Illumina	Amplicon and shotgun sequencing	Ranjan et al. 2016
Study on human gut microbiota	Illumina, Ion Torrent	Amplicon sequencing	Panek et al. 2018
Comparison of two platforms on complex microbiota composition using tandem variable 16S rRNA gene	Illumina, 454 Roche	Amplicon sequencing	Claesson et al. 2010
Evaluation of 16S rRNA amplicon sequencing	Illumina	V1–V3, V1–V8	Myer et al. 2016
Microbiome of ticks	Ion Torrent	V2, V3, V4, V67, V8 and V9	Sperling et al. 2017

diversity analysis (i) deep amplicons sequencing (targeted sequencing) (ii) whole genome sequencing (shotgun sequencing). Steps and work-flow of amplicon and shotgun sequencing are shown (Fig. 9).

Amplicon sequencing

It relies on sequencing of a molecular marker gene such as ribosomal RNA gene. 16S rRNA gene plays a crucial role in study of bacterial diversity, ecology, evolution and has become a standard marker for taxonomic allocation of wide range of taxa in phylogenetic and diversity analysis. The gene (approx. 1500 bp long) consists of nine hypervariable regions (V1–V9), each flanked by highly evolutionarily conserved regions. Several properties make this gene as unique and decidedly reliable because of the following: (i) they are evolutionarily conserved, making it a reliable molecular chronometer, (ii) the gene has diverged sufficiently that polymorphisms across their highly hypervariable regions, viz., V3, V3–V4 or V6 (Cai et al. 2013, Staley et al. 2013) enhance the resolution of closely related species into individual phylotypes (Hugerth and Andersson 2017), (iii) the hypervariable regions are flanked with conserved regions that enables the design of PCR primers and hybridization probes and (iv) this gene is weakly affected by horizontal gene transfer.

Unlike other diversity analysis methods, all NGS platforms are limited to a sequence of 350–500 bp in length. Both bioinformatics and empirical research have demonstrated that selection of appropriate hypervariable region is a crucial parameter in 16S rRNA gene based diversity studies. Further, in various environments, different studies suggested different justification on selection of primer set spanning V3–V4 regions (Cardoso et al. 2017), V1–V3 regions (Li et al. 2009), V3 region, V1–V2 regions and V4 region.

Diversity study utilizing 16S rRNA gene marker assessment is also known as "metagenetic sequencing" (Creer et al. 2010), "metasystematic sequencing" (Hajibabaei 2012), "metaprofiling" (Creer et al. 2016) and currently it is popularized with the term "DNA metabarcoding". Bacterial diversity can be assessed by

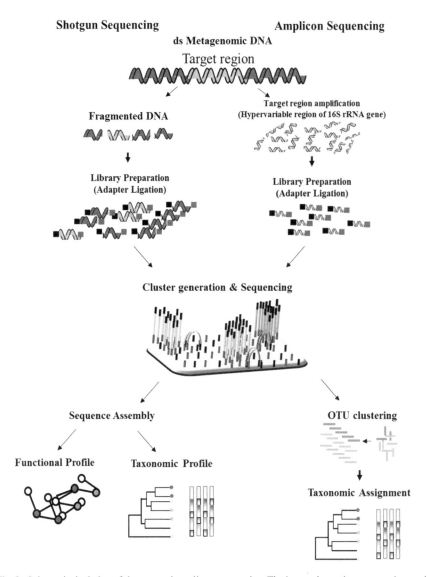

Fig. 9. Schematic depiction of shotgun and amplicon sequencing. The image shows the comparative work flow of whole metagenome shotgun and amplicon sequencing for diversity analysis.

exploiting the conserved regions of 16S rRNA gene with selected amplification of target sites. Effective combination of primer sets to target different hypervariable region facilitates bacterial identification even at genus and species level.

For bacterial community profiling through amplicon sequencing, generally the following common steps are used: (i) isolation of high-quality metagenomic DNA from source sample (ii) amplification of hypervariable region(s) of 16S rRNA gene using suitable primer set (iii) library preparation and sequencing (iv) analysis of sequences using bioinformatics tools including denoising, detection of chimeric sequences, and finally (v) OTU clustering and taxonomic assignment. Factors such as quality of metagenomic DNA, primer set used for amplification of hypervariable regions, purification, protocol used for library construction, sequencing platforms, sequencing depth and even database used for sequences analysis can significantly influence the diversity results. Amplicon/targeted sequencing of 16S rRNA gene has assisted analysis of bacterial diversity in all ecosystems including soil, marine sediment (Aravindraja et al. 2013), atmosphere (Serrano et al. 2018), marine biofilm (Rajeev et al. 2019), freshwater, caves, forests, gut microbiota, extreme habitats, viz., geothermal hot springs, acidophilic habitats, hyper-saline habitats. The amplicon-based bacterial diversity analysis has been also targeted in characterization of global microbial taxonomy and functional diversity in the Earth Microbiome Project.

The data generated by amplicon-based NGS has given a comprehensive information on environmental health and stability and paved way for understanding the health of an ecosystem. A considerable number of studies have been carried out on bacterial diversity variations with respect to environmental fluctuations such as effect of 2004 tsunami on bacterial community of marine, freshwater and brackish water in Thailand (Somboonna et al. 2014), effect of temperature fluctuation on bacterial diversity of amphibian skin (Longo and Zamudio 2017), climate warming on gut microbiota of an ectotherm (Bestion et al. 2017), and climatic variation on global soil bacterial diversity (Delgado-Baquerizo et al. 2016). In case of anthropogenic activities, analysis of heavy metals dominates the bacterial diversity studies. Recently, research targeting industrial metal contamination revealed the effects of bio-toxicity and bioaccumulation on ecosystem trophic levels especially on sediment microbiome (Jacquiod et al. 2018) in French River, mine tailings and Au-Ag accumulation (Kwon et al. 2015). Both these studies reported how bacterial diversity gets affected even at small scale because of heavy metal pollution. Apart from this, effect of other metals such as Mercury (Mahbub et al. 2017), lead, zinc and copper (Kou et al. 2018) has also been reported.

Host and habitat microbiome, their interactions and interacting diversity have skyrocketed recently, as the knowledge on microbiome and its diversity has increased rapidly through metacommunity sequencing. All eukaryotic organisms are chimeric creatures concealed with microbial cells. These microorganisms make up the individual's microbiome. Host and its habitat microbiome interactions are known to have impact on the metabolic, developmental and immunological status of the host. One such key stone study on human gut microbial function was performed using Illumina platform (Qin et al. 2010). Since then, the role of microbial diversity in microbiome of various hosts, viz., mammals (Zhu et al. 2011), insects (Jupatanakul et al. 2014, Vacheron et al. 2019), corals (Ainsworth et al. 2015, Thurber et al. 2009), reptiles (Pardo et al. 2017, Keenan et al. 2013), birds (Kreisinger et al. 2015, Jacob et al. 2014) has been explored. NGS plays a major role in divulging the host associated microbiome and its contribution towards the host survival. In clinical setting, amplicon-based studies of gut microbiota with respect to an individual's diet, surrounding environment, disease conditions and geography have been a topic of immense investigation that has established a correlation among diet, microbiota, host immune system and health conditions (Yatsunenko et al. 2012). Furthermore, reduction and/or fluctuation in structure of gut microbiome has been frequently associated with variety of diseases such as type II diabetes (Qin et al. 2012), and obesity (Turnbaugh et al. 2009).

Whole genome sequencing (WGS)

It is a shotgun approach that determines the entire genome sequence of an organism in a single run. This approach uses random primers to sequence overlapping regions of a complete genome or metagenome.

WGS sequencing is made possible with 'Shotgun sequencing'—an approach involving sequencing of long DNA strands by breaking up DNA sequences randomly into fragments. These fragments are sequenced and reassembled using computer program by finding the overlapping regions (Contig assembly). A typical shotgun sequencing comprises the following main steps: (i) sample collection and extraction of high quality metagenomic DNA (ii) library preparation of random fragmented DNA and sequencing (iii) preprocessing of sequencing reads and (iv) contig assembly and binning contigs (assembly of sequences into contigs, OTU clustering and taxonomic assignments to genus or till species level). Functional annotation and analysis can correspondingly be made with shotgun sequencing by employing a suitable software, for example PICRUST (Phylogenetic Investigation of Communities by Reconstruction of Unobserved States).

Amplicon sequencing that targets only specific molecular marker, fails to cover all the available diversity, while WGS conceals entire diversity with high genomic information and taxonomic resolution along with which the function can also be predicted. This is possible as WGS uses random primers and is not constrained by sequence conservation or with identification of primer binding sites of specific target.

Shotgun sequencing, similar to amplicon, has made contribution to various fields. WGS has made immense accomplishments and investigated intricate microbiomes such as Human gut microbiome, human skin flora, global ocean microbiome and outbreak of Shiga-toxigenic *Escherichia coli* (Huttenhower et al. 2012, Oh et al. 2014, Loman et al. 2013, Sunagawa et al. 2015). This approach has also been employed to decipher the endosymbiotic behavior of certain bacterial phyla (Brown et al. 2015), identification of bacterial species that can completely nitrify the ammonia (Daims et al. 2015), presence of antibiotic gene in commensal gut bacteria (Donia et al. 2014), and examination of strain level variations in gut microbiota after fecal microbiome transplantation (Li et al. 2016) and has aided in identification of several novel genes allied with osmoregulation and dormancy in desert bacterial community, genes associated with phytic acid utilization. In the field of clinical microbiology, the input of shotgun sequencing was firstly on the case of encephalitis (Wilson et al. 2014) and this paved the way to use WGS as a pathogen detection tool. Whole Genome Sequencing has also entered into the personal genome that resulted in sequencing the entire genome of an individual (Kim et al. 2009).

Although both amplicon-sequencing and whole genome sequencing have their own limitations, a considerable number of studies have been carried out (Table 3). The main advantage of WGS is that the identification of taxa can be more accurately defined even at species level and provide higher diversity than amplicon-sequencing (Tessler et al. 2017). Few limitations of WGS make it the least employed compared to amplicon sequencing, viz., the high cost, large quantity of data and extensive data analysis.

Third-generation sequencing

All NGS platforms are hampered with two major limitations: (i) generation of short read length that requires bioinformatic tools for assembly (ii) requires PCR amplification of target gene that itself may introduce biases. In Third-Generation Sequencing (TGS), instead of sequencing of PCR amplified products, single molecule can be sequenced and therefore termed as Single Molecule Sequencing (SMS). SMS approach supersedes SBS and SBL by (i) observing single molecule of DNA polymerase to synthesize single molecule of DNA template that overcomes PCR biases, (ii) Nanopore sequencing technology facilitating the detection of individual bases when it passes through the nanopore that surmounts wash-and-scan system and finally (iii) sophisticated imaging system with advanced microscopy techniques that can directly see the integration of single nucleotide by solitary DNA polymerase defeating the need of fluorophore/terminators, scanning and imaging. TGS employs Single Molecule Real-Time (SMRT) approach to sequence DNA, which hires Zero-Mode Waveguide (ZMW) technology offering single base detection from a pile of bases carrying light signals. ZMW—a pore in 100 nm metal film, selectively detects the visible light of ~ 30 nm than 600 nm (actual wavelength of visible light) by simple diffusion of visible light through its pore that gets decayed to 30 nm. DNA polymerase attached to the metal film through biotin-streptavidin interaction aids the incorporation of correct nucleotide, thereby facilitating sequencing (Schadt et al. 2010). Beyond bacterial diversity identification, the raise of TGS gives an unprecedented picture of complex biological analysis with more accuracy and with great predictive power. Current technologies in TGS include few commercialized platforms such as SMAT by Pacific Biosciences and Oxford Nanopore's MinION.

Nanopore sequencing

Currently, Oxfords Nanopore's MinION anticipates the era of third generation sequencing technology. The core concept of sequencing through Nanopore's MinION is based on disruption in voltage of a tiny biopore (nano-scaled pore that facilitates ion exchange in biological membranes) during particle movement. Single stranded DNA is electrophoretically passed through the nanopore of α-haemolysin (αHL) protein in a controlled manner using motor protein. αHL (isolated from *Staphylococcus aureus*), a 33kD protein, self-assembles to form heptameric transmembrane channel for a molecule to pass through. As the single stranded DNA is translocated across the membrane pore, changes in current (current shifts) are recorded in real-time and graphically represented as squiggle plot. Alternative to αHL, the MspA pore and the CsgG pore are currently used in Oxfords Nanopore technology. Sensing region of a pore determines the accuracy of base determination. The MspA pore and the CsgG pore have shorter sensing region that senses even smaller number of nucleotides, leading to more precise base detection. The advantage offered by Oxfords Nanopore technology is its USB size, palm portable DNA sequencer with higher read length and shorter run time (> 5 kbp read length with speed of 1 bp/ns) to sequence. Moreover, the non-usage of PCR reduces the quantity of starting material and PCR-based biases. Oxfords Nanopore technology also sequences both strands of DNA that offers less error rate.

Concluding remarks and future prospective

Planet Earth harbors a large spectrum of microorganisms. For nearly 250 years, microbiologists have been struggling to catalog and understand their interaction as well as functionality in natural environment as the efforts were majorly hampered by technical development. Advances in DNA sequencing technologies such as high-throughput next-generation sequencing and progress in computational methods have made a tremendous upsurge in biological research.

As the advancements in sequencing platforms progress, the affordability and feasibility of examining total metagenome of any habitat are becoming reality. Diversity and functional analysis of habitat's microbiome are attaining its peak of precision allowing best resolution of taxa. NGS has allowed to understand the intricate networks between the human society, environment and microbial communities. Today, we are in a position to define an environment from its microbial footprints. From the human body to the surrounding environment, the planet is filled with microbes and it is clear that personal microbiome sheds everywhere we move. Hence, one is curious to ask whether it is possible to design and/or rebuilt our own microbiome to keep ourselves healthy. Or can we make all the humans as gnotobiotic (germ-free)? Can we trace the lifetime of any individual through sequencing technologies? Answer to all these questions is possible through developing sequencing platforms. The upcoming sequencing generation developments will shed more light on the unculturable microorganisms and its biochemical advantages.

Despite the existing advancements in NGS technologies, the method progression and bioinformatic tools to analyze data are still in their infancy. Although the advanced new generation sequencing methods have stepped into the field of microbial diversity, Illumina is highly preferred over other NGS platforms because of its high sequencing efficiency with least cost and well established data analysis pipelines. Moreover, it is certain that new sequencing technologies will evolve rapidly, resulting in powerful and cheaper platforms. It is in the hands of researchers to have informed choices on NGS platforms to fulfill their research objectives.

In this chapter, we discussed comprehensive overview of generation of sequencing technologies commenced by DNA sequencing history followed by comparison of three sequencing generations, various platforms of NGS and application in microbial diversity analyses. Though there are associated challenges in NGS such as intrinsic sequencing errors, PCR biases, counting the artifacts, difficulties in storing and analyzing the data generated, understanding and appreciation with exploitation of current methodologies will provide new insights to various fields of biology.

Acknowledgements

MR gratefully thank the financial support provided by RUSA Phase 2.0 [F.24-51/2014-U, Policy (TN Multi-Gen), Dept. of Edn, GoI] in the form of Ph.D. Fellowship. Authors sincerely acknowledge the computational and bioinformatics facility provided by the Bioinformatics Infrastructure Facility (funded by DBT, GOI; File No. BT/BI/25/012/2012, BIF). The authors also thankfully acknowledge DST-FIST (Grant No. SR/FST/LSI-639/2015 (C)), UGC-SAP (Grant No. F.5-1/2018/ DRS-II (SAP-II)) and DST-PURSE (Grant No. SR/PURSE Phase 2/38 (G)) for providing instrumentation facilities.

References

Adamiak, J., Otlewska, A., Tafer, H., Lopandic, K., Gutarowska, B., Sterflinger, K. et al. 2018. First evaluation of the microbiome of built cultural heritage by using the Ion Torrent next generation sequencing platform. International Biodeterioration & Biodegradation 131: 11–18.

Aguirre, M., Abad, D., Albaina, A., Cralle, L., Goñi-Urriza, M. S., Estonba, A. et al. 2017. Unraveling the environmental and anthropogenic drivers of bacterial community changes in the Estuary of Bilbao and its tributaries. PLoS One 12(6): e0178755.

Ainsworth, T. D., Krause, L., Bridge, T., Torda, G., Raina, J. B., Zakrzewski, M. et al. 2015. The coral core microbiome identifies rare bacterial taxa as ubiquitous endosymbionts. The ISME Journal 9(10): 2261.

Allen, E. E. and Banfield, J. F. 2005. Community genomics in microbial ecology and evolution. Nature Reviews Microbiology 3(6): 489.

Aravindraja, C., Viszwapriya, D. and Pandian, S. K. 2013. Ultradeep 16S rRNA sequencing analysis of geographically similar but diverse unexplored marine samples reveal varied bacterial community composition. PLoS One 8(10): e76724.

Besser, J., Carleton, H. A., Gerner-Smidt, P., Lindsey, R. L. and Trees, E. 2018. Next-generation sequencing technologies and their application to the study and control of bacterial infections. Clinical Microbiology and Infection 24(4): 335–341.

Bestion, E., Jacob, S., Zinger, L., Di Gesu, L., Richard, M., White, J. et al. 2017. Climate warming reduces gut microbiota diversity in a vertebrate ectotherm. Nature Ecology and Evolution 1(6): 0161.

Brown, C. T., Hug, L. A., Thomas, B. C., Sharon, I., Castelle, C. J., Singh, A. et al. 2015. Unusual biology across a group comprising more than 15% of domain bacteria. Nature 523(7559): 208.

Cai, L., Ye, L., Tong, A. H. Y., Lok, S. and Zhang, T. 2013. Biased diversity metrics revealed by bacterial 16S pyrotags derived from different primer sets. PLoS One 8(1): e53649.

Cardoso, D. C., Sandionigi, A., Cretoiu, M. S., Casiraghi, M., Stal, L. and Bolhuis, H. 2017. Comparison of the active and resident community of a coastal microbial mat. Scientific Reports 7(1): 2969.

Celikkol-Aydin, S., Gaylarde, C. C., Lee, T., Melchers, R. E., Witt, D. L. and Beech, I. B. 2016. 16S rRNA gene profiling of planktonic and biofilm microbial populations in the Gulf of Guinea using Illumina NGS. Marine Environmental Research 122: 105–112.

Claesson, M. J., Wang, Q., O'sullivan, O., Greene-Diniz, R., Cole, J. R., Ross, R. P. et al. 2010. Comparison of two next-generation sequencing technologies for resolving highly complex microbiota composition using tandem variable 16S rRNA gene regions. Nucleic Acids Research 38(22): e200–e200.

Clooney, A. G., Fouhy, F., Sleator, R. D., O'Driscoll, A., Stanton, C., Cotter, P. D. et al. 2016. Comparing apples and oranges?: next generation sequencing and its impact on microbiome analysis. PLoS One 11(2): e0148028.

Creer, S., Fonseca, V. G., Porazinska, D. L., Giblin-Davis, R. M., Sung, W., Power, D. M. et al. 2010. Ultrasequencing of the meiofaunal biosphere: practice, pitfalls and promises. Molecular Ecology 19: 4–20.

Creer, S., Deiner, K., Frey, S., Porazinska, D., Taberlet, P., Thomas, W. K. et al. 2016. The ecologist's field guide to sequence-based identification of biodiversity. Methods in Ecology and Evolution 7(9): 1008–1018.

Cremonesi, P., Conte, G., Severgnini, M., Turri, F., Monni, A., Capra, E. et al. 2018. Evaluation of the effects of different diets on microbiome diversity and fatty acid composition of rumen liquor in dairy goat. Animal 12(9): 1856–1866.

Daims, H., Lebedeva, E. V., Pjevac, P., Han, P., Herbold, C., Albertsen, M. et al. 2015. Complete nitrification by Nitrospira bacteria. Nature 528(7583): 504.

De Mandal, S., Panda, A. K., Bisht, S. S. and Kumar, N. S. 2015. Microbial ecology in the era of next generation sequencing. Next Generation: Sequencing & Applications S: 1–2.

Delgado-Baquerizo, M., Maestre, F. T., Reich, P. B., Trivedi, P., Osanai, Y., Liu, Y. R. et al. 2016. Carbon content and climate variability drive global soil bacterial diversity patterns. Ecological Monographs 86(3): 373–390.

Donia, M. S., Cimermancic, P., Schulze, C. J., Brown, L. C. W., Martin, J., Mitreva, M. et al. 2014. A systematic analysis of biosynthetic gene clusters in the human microbiome reveals a common family of antibiotics. Cell 158(6): 1402–1414.

Drmanac, R., Sparks, A. B., Callow, M. J., Halpern, A. L., Burns, N. L., Kermani, B. G. et al. 2010. Human genome sequencing using unchained base reads on self-assembling DNA nanoarrays. Science 327(5961): 78–81.

Escalante, A. E., Barbolla, L. J., Ramírez-Barahona, S. and Eguiarte, L. E. 2014. The study of biodiversity in the era of massive sequencing. Mexican Journal of Biodiversity 85(4): 1249–1264.

Feng, B. W., Li, X. R., Wang, J. H., Hu, Z. Y., Meng, H., Xiang, L. Y. et al. 2009. Bacterial diversity of water and sediment in the Changjiang estuary and coastal area of the East China Sea. FEMS Microbiology Ecology 70(2): 236–248.

Fierer, N., Leff, J. W., Adams, B. J., Nielsen, U. N., Bates, S. T., Lauber, C. L. et al. 2012. Cross-biome metagenomic analyses of soil microbial communities and their functional attributes. Proceedings of the National Academy of Sciences of the United States of America 109(52): 21390–21395.

Fleischmann, R. D., Adams, M. D., White, O., Clayton, R. A., Kirkness, E. F., Kerlavage, A. R. et al. 1995. Whole-genome random sequencing and assembly of Haemophilus influenzae Rd. Science 269(5223): 496–512.

Gilbert, W. and Maxam, A. 1973. The nucleotide sequence of the lac operator. Proceedings of the National Academy of Sciences of the United States of America 70(12): 3581–3584.

Hajibabaei, M. 2012. The golden age of DNA metasystematics. Trends in Genetics 28(11): 535–537.

Handelsman, J. 2004. Metagenomics: application of genomics to uncultured microorganisms. Microbiology and Molecular Biology Reviews 68(4): 669–685.

Hicks, N., Liu, X., Gregory, R., Kenny, J., Lucaci, A., Lenzi, L. et al. 2018. Temperature driven changes in benthic bacterial diversity influences biogeochemical cycling in coastal sediments. Frontiers in Microbiology 9: 1730.

Hou, L., Zhou, Q., Wu, Q., Gu, Q., Sun, M. and Zhang, J. 2018. Spatiotemporal changes in bacterial community and microbial activity in a full-scale drinking water treatment plant. Science of the Total Environment 625: 449–459.

Hugerth, L. W. and Andersson, A. F. 2017. Analysing microbial community composition through amplicon sequencing: from sampling to hypothesis testing. Frontiers in Microbiology 8: 1561.

Hunkapiller, T., Kaiser, R. J., Koop, B. F. and Hood, L. 1991. Large-scale and automated DNA sequence determination. Science 254(5028): 59–67.

Huttenhower, C., Gevers, D., Knight, R., Abubucker, S., Badger, J. H., Chinwalla, A. T. et al. 2012. Structure, function and diversity of the healthy human microbiome. Nature 486(7402): 207.

Illeghems, K., De Vuyst, L., Papalexandratou, Z. and Weckx, S. 2012. Phylogenetic analysis of a spontaneous cocoa bean fermentation metagenome reveals new insights into its bacterial and fungal community diversity. PLoS One 7(5): e38040.

Jacob, S., Immer, A., Leclaire, S., Parthuisot, N., Ducamp, C., Espinasse, G. et al. 2014. Uropygial gland size and composition varies according to experimentally modified microbiome in Great tits. BMC Evolutionary Biology 14(1): 134.

Jacquiod, S., Cyriaque, V., Riber, L., Al-Soud, W. A., Gillan, D. C., Wattiez, R. et al. 2018. Long-term industrial metal contamination unexpectedly shaped diversity and activity response of sediment microbiome. Journal of Hazardous Materials 344: 299–307.

Jeffries, T. C., Schmitz Fontes, M. L., Harrison, D. P., Van-Dongen-Vogels, V., Eyre, B. D., Ralph, P. J. et al. 2016. Bacterioplankton dynamics within a large anthropogenically impacted urban estuary. Frontiers in Microbiology 6: 1438.

Jupatanakul, N., Sim, S. and Dimopoulos, G. 2014. The insect microbiome modulates vector competence for arboviruses. Viruses 6(11): 4294–4313.

Kaczmarczyk, A., Kucharczyk, H., Kucharczyk, M., Kapusta, P., Sell, J. and Zielińska, S. 2018. First insight into microbiome profile of fungivorous thrips Hoplothrips carpathicus (Insecta: Thysanoptera) at different developmental stages: molecular evidence of Wolbachia endosymbiosis. Scientific Reports 8(1): 14376.

Karlsson, F. H., Tremaroli, V., Nookaew, I., Bergström, G., Behre, C. J., Fagerberg, B. et al. 2013. Gut metagenome in European women with normal, impaired and diabetic glucose control. Nature 498(7452): 99.

Kchouk, M., Gibrat, J. F. and Elloumi, M. 2017. Generations of sequencing technologies: From first to next generation. Biology and Medicine 9: 3.

Keenan, S. W., Engel, A. S. and Elsey, R. M. 2013. The alligator gut microbiome and implications for archosaur symbioses. Scientific Reports 3: 2877.

Kim, J. I., Ju, Y. S., Park, H., Kim, S., Lee, S., Yi, J. H. et al. 2009. A highly annotated whole-genome sequence of a Korean individual. Nature 460(7258): 1011.

Kou, S., Vincent, G., Gonzalez, E., Pitre, F. E., Labrecque, M. and Brereton, N. J. 2018. The response of a 16S ribosomal RNA gene fragment amplified community to lead, zinc, and copper pollution in a shanghai field trial. Frontiers in Microbiology 9: 366.

Kreisinger, J., Čížková, D., Kropáčková, L. and Albrecht, T. 2015. Cloacal microbiome structure in a long-distance migratory bird assessed using deep 16sRNA pyrosequencing. PLoS One 10(9): 0137401.

Kwon, M. J., Yang, J. S., Lee, S., Lee, G., Ham, B., Boyanov, M. I. et al. 2015. Geochemical characteristics and microbial community composition in toxic metal-rich sediments contaminated with Au–Ag mine tailings. Journal of Hazardous Materials 296: 147–157.

Lawes, J. C., Neilan, B. A., Brown, M. V., Clark, G. F. and Johnston, E. L. 2016. Elevated nutrients change bacterial community composition and connectivity: high throughput sequencing of young marine biofilms. Biofouling 32(1): 57–69.

Leite, A. M., Mayo, B., Rachid, C. T., Peixoto, R. S., Silva, J. T., Paschoalin, V. M. F. et al. 2012. Assessment of the microbial diversity of Brazilian kefir grains by PCR-DGGE and pyrosequencing analysis. Food Microbiology 31(2): 215–221.

Li, S. S., Zhu, A., Benes, V., Costea, P. I., Hercog, R., Hildebrand, F. et al. 2016. Durable coexistence of donor and recipient strains after fecal microbiota transplantation. Science 352(6285): 586–589.

Liu, M. B., Xu, S. R., He, Y., Deng, G. H., Sheng, H. F., Huang, X. M. et al. 2013. Diverse vaginal microbiomes in reproductive-age women with vulvovaginal candidiasis. PLoS One 8(11): e79812.

Loman, N. J., Constantinidou, C., Christner, M., Rohde, H., Chan, J.Z.M., Quick, J. et al. 2013. A culture-independent sequence-based metagenomics approach to the investigation of an outbreak of Shiga-toxigenic *Escherichia coli* O104:H4. Jama 309(14): 1502–1510.

Longo, A. V. and Zamudio, K. R. 2017. Temperature variation, bacterial diversity and fungal infection dynamics in the amphibian skin. Molecular Ecology 26(18): 4787–4797.

Ma, L., Mao, G., Liu, J., Gao, G., Zou, C., Bartlam, M. G. et al. 2016. Spatial-temporal changes of bacterioplankton community along an exhorheic river. Frontiers in Microbiology 7: 250.

Mahbub, K. R., Subashchandrabose, S. R., Krishnan, K., Naidu, R. and Megharaj, M. 2017. Mercury alters the bacterial community structure and diversity in soil even at concentrations lower than the guideline values. Applied Microbiology and Biotechnology 101(5): 2163–2175.

Mardis, E. R. 2008. Next-generation DNA sequencing methods. Annual Review of Genomics and Human Genetics 9: 387–402.

Mardis, E. R. 2008. The impact of next-generation sequencing technology on genetics. Trends in Genetics 24(3): 133–141.

Marri, P. R., Stern, D. A., Wright, A. L., Billheimer, D. and Martinez, F. D. 2013. Asthma-associated differences in microbial composition of induced sputum. The Journal of Allergy and Clinical Immunology: In Practice 131(2): 346–352.

Masoud, W., Vogensen, F. K., Lillevang, S., Al-Soud, W. A., Sørensen, S. J. and Jakobsen, M. 2012. The fate of indigenous microbiota, starter cultures, *Escherichia coli*, *Listeria innocua* and *Staphylococcus aureus* in Danish raw milk and cheeses determined by pyrosequencing and quantitative real time (qRT)-PCR. International Journal of Food Microbiology 153(1-2): 192–202.

Maxam, A. M. and Gilbert, W. 1977. A new method for sequencing DNA. Proceedings of the National Academy of Sciences of the United States of America 74(2): 560–564.

McCormack, J. E., Hird, S. M., Zellmer, A. J., Carstens, B. C. and Brumfield, R. T. 2013. Applications of next-generation sequencing to phylogeography and phylogenetics. Molecular Phylogenetics and Evolution 66(2): 526–538.

Metzker, M. L. 2010. Sequencing technologies—the next generation. Nature Reviews Genetics 11(1): 31.

Mitra, A., MacIntyre, D. A., Lee, Y. S., Smith, A., Marchesi, J. R., Lehne, B. et al. 2015. Cervical intraepithelial neoplasia disease progression is associated with increased vaginal microbiome diversity. Scientific Reports 5: 16865.

Myer, P. R., Kim, M., Freetly, H. C. and Smith, T. P. 2016. Evaluation of 16S rRNA amplicon sequencing using two next-generation sequencing technologies for phylogenetic analysis of the rumen bacterial community in steers. Journal of Microbiological Methods 127: 132–140.

Oh, J., Byrd, A. L., Deming, C., Conlan, S., Barnabas, B., Blakesley, R. et al. 2014. Biogeography and individuality shape function in the human skin metagenome. Nature 514(7520): 59.

Pan, D. and Yu, Z. 2014. Intestinal microbiome of poultry and its interaction with host and diet. Gut Microbes 5(1): 108–119.

Panek, M., Paljetak, H. Č., Barešić, A., Perić, M., Matijašić, M., Lojkić, I. et al. 2018. Methodology challenges in studying human gut microbiota–effects of collection, storage, DNA extraction and next generation sequencing technologies. Scientific Reports 8(1): 5143.

Pedersen, K. 2000. Exploration of deep intraterrestrial microbial life: current perspectives. FEMS Microbiology Letters 185(1): 9–16.

Poirier, S., Rué, O., Peguilhan, R., Coeuret, G., Zagorec, M., Champomier-Vergès, M. C. et al. 2018. Deciphering intra-species bacterial diversity of meat and seafood spoilage microbiota using gyrB amplicon sequencing: A comparative analysis with 16S rDNA V3-V4 amplicon sequencing. PLoS One 13(9): e0204629.

Poretsky, R., Rodriguez-R, L. M., Luo, C., Tsementzi, D. and Konstantinidis, K. T. 2014. Strengths and limitations of 16S rRNA gene amplicon sequencing in revealing temporal microbial community dynamics. PLoS One 9(4): e93827.

Prado-Irwin, S. R., Bird, A. K., Zink, A. G. and Vredenburg, V. T. 2017. Intraspecific variation in the skin-associated microbiome of a terrestrial salamander. Microbial Ecology 74(3): 745–756.

Qin, J., Li, R., Raes, J., Arumugam, M., Burgdorf, K. S., Manichanh, C. et al. 2010. A human gut microbial gene catalogue established by metagenomic sequencing. Nature 464(7285): 59.

Qin, J., Li, Y., Cai, Z., Li, S., Zhu, J., Zhang, F. et al. 2012. A metagenome-wide association study of gut microbiota in type 2 diabetes. Nature 490(7418): 55.

Quero, G. M., Cassin, D., Botter, M., Perini, L. and Luna, G. M. 2015. Patterns of benthic bacterial diversity in coastal areas contaminated by heavy metals, polycyclic aromatic hydrocarbons (PAHs) and polychlorinated biphenyls (PCBs). Frontiers in Microbiology 6: 1053.

Rajeev, M., Sushmitha, T. J., Toleti, S. R. and Pandian, S. K. 2019. Culture dependent and independent analysis and appraisal of early stage biofilm-forming bacterial community composition in the Southern coastal seawater of India. Science of the Total Environment 666: 308–320.

Ranjan, R., Rani, A., Metwally, A., McGee, H. S. and Perkins, D. L. 2016. Analysis of the microbiome: advantages of whole genome shotgun versus 16S amplicon sequencing. Biochemical and Biophysical Research Communications 469(4): 967–977.

Rôças, I. N., Alves, F. R., Rachid, C. T., Lima, K. C., Assunção, I. V., Gomes, P. N. et al. 2016. Microbiome of deep dentinal caries lesions in teeth with symptomatic irreversible pulpitis. PLoS One 11(5): e0154653.

Sanger, F., Nicklen, S. and Coulson, A. R. 1977. DNA sequencing with chain-terminating inhibitors. Proceedings of the National Academy of Sciences of the United States of America 74(12): 5463–5467.

Sanger, F. 1988. Sequences, sequences, and sequences. Annual Review of Biochemistry 57(1): 1–29.

Schadt, E. E., Turner, S. and Kasarskis, A. 2010. A window into third-generation sequencing. Human Molecular Genetics 19(R2): R227–R240.

Serrano-Silva, N. and Calderón-Ezquerro, M.C. 2018. Metagenomic survey of bacterial diversity in the atmosphere of Mexico City using different sampling methods. Environmental Pollution 235: 20–29.

Shendure, J. and Ji, H. 2008. Next-generation DNA sequencing. Nature Biotechnology 26(10):1135.

Shendure, J., Balasubramanian, S., Church, G. M., Gilbert, W., Rogers, J., Schloss, J. A. et al. 2017. DNA sequencing at 40: past, present and future. Nature 550(7676): 345.

Smith, L. M., Fung, S., Hunkapiller, M. W., Hunkapiller, T. J. and Hood, L. E. 1985. The synthesis of oligonucleotides containing an aliphatic amino group at the 5′ terminus: synthesis of fluorescent DNA primers for use in DNA sequence analysis. Nucleic Acids Research 13(7): 2399–2412.

Sogin, M. L., Morrison, H. G., Huber, J. A., Welch, D. M., Huse, S. M., Neal, P. R. et al. 2006. Microbial diversity in the deep sea and the underexplored "rare biosphere". Proceedings of the National Academy of Sciences of the United States of America 103(32): 12115–12120.

Somboonna, N., Wilantho, A., Jankaew, K., Assawamakin, A., Sangsrakru, D., Tangphatsornruang, S. et al. 2014. Microbial ecology of Thailand tsunami and non-tsunami affected terrestrials. PLoS One 9(4): e94236.

Song, H., Li, Z., Du, B., Wang, G. and Ding, Y. 2012. Bacterial communities in sediments of the shallow Lake Dongping in China. Journal of Applied Microbiology 112(1): 79–89.

Sperling, J. L., Silva-Brandao, K. L., Brandao, M. M., Lloyd, V. K., Dang, S., Davis, C. S. et al. 2017. Comparison of bacterial 16S rRNA variable regions for microbiome surveys of ticks. Ticks and Tick-borne Diseases 8(4): 453–461.

Staley, C., Unno, T., Gould, T. J., Jarvis, B., Phillips, J., Cotner, J. B. et al. 2013. Application of Illumina next-generation sequencing to characterize the bacterial community of the Upper Mississippi River. Journal of Applied Microbiology 115(5): 1147–1158.

Sunagawa, S., Coelho, L. P., Chaffron, S., Kultima, J. R., Labadie, K., Salazar, G. et al. 2015. Structure and function of the global ocean microbiome. Science 348(6237): 1261359.

Swerdlow, H. and Gesteland, R. 1990. Capillary gel electrophoresis for rapid, high resolution DNA sequencing. Nucleic Acids Research 18(6): 1415–1419.

Tessler, M., Neumann, J. S., Afshinnekoo, E., Pineda, M., Hersch, R., Velho, L. F. M. et al. 2017. Large-scale differences in microbial biodiversity discovery between 16S amplicon and shotgun sequencing. Scientific Reports 7(1): p.6589.

The, C. E. S. C. 1998. Genome sequence of the nematode *C. elegans*: a platform for investigating biology. Science 282(5396): 2–012.

Thurber, R. V., Willner-Hall, D., Rodriguez-Mueller, B., Desnues, C., Edwards, R. A., Angly, F. et al. 2009. Metagenomic analysis of stressed coral holobionts. Environmental Microbiology 11(8): 2148–2163.

Turnbaugh, P. J., Hamady, M., Yatsunenko, T., Cantarel, B. L., Duncan, A., Ley, R. E. et al. 2009. A core gut microbiome in obese and lean twins. Nature 457(7228): 480.

Vacheron, J., Péchy-Tarr, M., Brochet, S., Heiman, C. M., Stojiljkovic, M., Maurhofer, M. et al. 2019. T6SS contributes to gut microbiome invasion and killing of an herbivorous pest insect by plant-beneficial *Pseudomonas protegens*. The ISME Journal 1.

Vasconcelos, E. J., Billeter, S. A., Jett, L. A., Meinersmann, R. J., Barr, M. C., Diniz, P. P. et al. 2018. Assessing Cat Flea microbiomes in Northern and Southern California by 16S rRNA Next-Generation sequencing. Vector-Borne and Zoonotic Diseases 18(9): 491–499.

Venter, J. C., Adams, M. D., Myers, E. W., Li, P. W., Mural, R. J., Sutton, G. G et al. 2001. The sequence of the human genome. Science 291(5507): 1304–1351.

Watson, J. D. and Crick, F. H. 1953. Molecular structure of nucleic acids. Nature 171(4356): 737–738.

Wilson, M. R., Naccache, S. N., Samayoa, E., Biagtan, M., Bashir, H., Yu, G. et al. 2014. Actionable diagnosis of neuroleptospirosis by next-generation sequencing. The New England Journal of Medicine 370(25): 2408–2417.